Young-Onset GI Cancer

Editors

Irit Ben Aharon
Savio George Barreto

MDPI • Basel • Beijing • Wuhan • Barcelona • Belgrade • Manchester • Tokyo • Cluj • Tianjin

Editors
Irit Ben Aharon
Division of Oncology,
Rambam Health Care Center
Haifa
Israel

Savio George Barreto
Flinders University
Adelaide
Australia

Editorial Office
MDPI
St. Alban-Anlage 66
4052 Basel, Switzerland

This is a reprint of articles from the Special Issue published online in the open access journal *Cancers* (ISSN 2072-6694) (available at: https://www.mdpi.com/journal/cancers/special_issues/Y-OGI).

For citation purposes, cite each article independently as indicated on the article page online and as indicated below:

LastName, A.A.; LastName, B.B.; LastName, C.C. Article Title. *Journal Name* **Year**, *Volume Number*, Page Range.

ISBN 978-3-0365-7964-1 (Hbk)
ISBN 978-3-0365-7965-8 (PDF)

Contents

cancers

MDPI

Editorial

Young-Onset Cancers—Early Steps in the Right Direction

Savio George Barreto [1,2,3,*] and Irit Ben-Aharon [4,5]

1 College of Medicine and Public Health, Flinders University, Adelaide, SA 5042, Australia
2 Division of Surgery and Perioperative Medicine, Flinders Medical Center, Bedford Park, Adelaide, SA 5042, Australia
3 Department of Surgery, Flinders Medical Centre, Bedford Park, Adelaide, SA 5042, Australia
4 Division of Oncology, Rambam Health Care Center, Haifa 3109601, Israel; iritbenaharon@gmail.com
5 Rappaport Faculty of Medicine, Technion, Haifa 3200003, Israel
* Correspondence: georgebarreto@yahoo.com or savio.barreto@flinders.edu.au; Tel.: +61-8-82045511

Citation: Barreto, S.G.; Ben-Aharon, I. Young-Onset Cancers—Early Steps in the Right Direction. *Cancers* **2023**, *15*, 2599. https://doi.org/10.3390/cancers15092599

Received: 23 April 2023
Accepted: 26 April 2023
Published: 4 May 2023

The global incidence of young-onset (YO) cancer is on the rise. Determining the cause of this disturbing trend has been listed as one of the Grand Cancer Challenges [1]. The incidence of YO cancers in the gastrointestinal tract are on the rise in the past two to three decades, in the United States, with YO colorectal cancer increasing across the world [2]. To note, incidence rates of some countries may be partly due to the lack of a strict cancer registry of every documented case, and, therefore, while there is a lot of work to be done to address the problem, the first steps include advocacy and the need to encourage reporting of the burden of the problem. This may help identify trends that could guide focused research into the etiology. This Special Issue hosted by *Cancers* entitled "Young-onset GI (gastrointestinal) cancer" presented clinicians and researchers around the world with a platform to submit their research focused on YO cancer.

This Special Issue boasts seven peer-reviewed publications, including three from Europe and two each from the United States and Australia. ***La Pelusa and colleagues*** provided important insights into the burden of early-onset gastric cancer (EOGC) from the United States [3]. Their findings highlight an increased likelihood of EOGC affecting female patients and individuals who identified as Asian/Pacific Islander, African American, and Hispanic. They also note that patients with EOGC are more likely to be uninsured and to present with stage IV disease compared with their older counterparts. The variability in cancer care (surgery and chemotherapy) delivery presents opportunities for intervention if we wish to improve survival within this subset of patients. ***Schell and Shepherdson and their colleagues*** from Australia provided compelling data (over two manuscripts) on the trends in YO GI (oesophagus, stomach, pancreas, colon, and rectum) adenocarcinomas from South Australia and the Northern Territory (of Australia) over the last 28 years [4,5]. Although the rising trends in YO GI adenocarcinomas in South Australia, especially amongst males, appear to attract one's attention, the existing high incidence for all YO GI adenocarcinomas in the Northern Territory, despite being unchanged over the study period, signals a worrying statistic that certainly warrants further investigation. The incidence rates noted for pancreatic cancer mirror the values seen in younger Black and Hispanic women in the United States [6]. ***Shepherdson* et al.** also addressed the incidence and survival rates amongst the Indigenous peoples (Aboriginal and Torres Strait Islander peoples) of Australia living in the Northern Territory [5]. The significantly lower survival compared with non-Indigenous peoples highlights an important area for health advocacy and the need for culturally safe Indigenous community-focused programs aimed at early detection and patient-centered management of GI adenocarcinomas. ***Ten Kate and colleagues*** from the Netherlands and Finland provide preliminary evidence on the susceptibility of the oesophageal epithelial homeostasis to acidic disturbances in individuals born with oesophageal atresia linking this observation to the increased propensity of this cohort of patients to develop early-onset Barrett's oesophagus [7]. ***Pocurull and colleagues*** from Spain performed a germinal genetic analysis on patients with EOGC [8]. They note that familial aggregation was observed in only 15% of cases, whilst a germline mutation was found in 25% of patients tested with clinical criteria. Their findings are important in terms of highlighting the genetic heterogeneity of EOGC thereby reinforcing the need for accurate genetic counseling as well as enhancing the emerging use of multigene panels.

The issue also features two review articles by ***Castle and colleagues*** [9] on paediatric neuroendocrine neoplasms and ***Dudzisz-Śledź and colleagues*** on the treatment of gastrointestinal stromal tumours in younger patients [10].

The aetiology of YO cancers remains to be clarified. Ben-Aharon et al. [11] recently provided an up-to-date review of early-onset GI cancers. Due to the fact the majority of these cases are sporadic, the aetiological factors imply a key role in environmental factors. If this is so, then why are we seeing a changing trend only amongst the young over the last few decades? The PELICan hypothesis [12,13] may help reconcile these differential effects of the same triggers (for carcinogenesis) on different individuals. Another study recently published in Nature [14] provides insight into the germline mutation rates in vertebrates, including humans. The rising trend in the incidence of YO cancers in males may be linked to the finding of Bergeron et al. [14] that per-generation mutation rates are much higher in the males of a species. So, how then can we explain the higher than usual YO cancer rates noted in younger females only within some racialised groups [3,6]? Shirazi and Rosinger [15] determined that non-Hispanic (NH) African American and Hispanic girls have a significantly lower age of menarche by about 4.3 (SE = 0.08, $p < 0.001$), and 3.2 months (SE = 0.09, $p < 0.001$), respectively, relative to NH white girls. Bergeron et al. [14] found that age at maturity and species-level fecundity are the key life-history traits affecting germline mutation variation among species. These hypotheses may explain the differential effect that aetiological factors may have on individuals to increase their risk of developing YO cancers.

Clearly, there is work to be done to improve the early detection and multi-disciplinary management of YO cancers. We remain hopeful that this Special Issue in *Cancers* will serve its purpose of advocating for action because our young people need our help.

Funding: Research was supported as follows: ***SGB support***–Flinders Foundation grant: 49358025. Pankind 21.R7.INV.CB.UOSA.6.2. NHMRC Ideas grant GNT2021009.

Conflicts of Interest: The authors declare no conflict of interest.

References

1. Nine New Challenges. Available online: https://cancergrandchallenges.org/challenges/early-onset-cancers (accessed on 23 April 2023).
2. Bray, F.; Ferlay, J.; Soerjomataram, I.; Siegel, R.L.; Torre, L.A.; Jemal, A. Global Cancer Statistics 2018: Globocan Estimates of Incidence and Mortality Worldwide for 36 Cancers in 185 Countries. *CA Cancer J. Clin.* **2018**, *68*, 394–424. [CrossRef]
3. LaPelusa, M.; Shen, C.; Gillaspie, E.A.; Cann, C.; Lambright, E.; Chakravarthy, A.B.; Gibson, M.K.; Eng, C. Variation in Treatment Patterns of Patients with Early-Onset Gastric Cancer. *Cancers* **2022**, *14*, 3633. [CrossRef] [PubMed]
4. Schell, D.; Ullah, S.; Brooke-Smith, M.E.; Hollington, P.; Yeow, M.; Karapetis, C.S.; Watson, D.I.; Pandol, S.; Roberts, C.T.; Barreto, S.G. Gastrointestinal Adenocarcinoma Incidence and Survival Trends in South Australia, 1990–2017. *Cancers* **2022**, *14*, 275. [CrossRef] [PubMed]
5. Shepherdson, M.; Leemaqz, S.; Singh, G.; Ryder, C.; Ullah, S.; Canuto, K.; Barreto, S.G. Young-Onset Gastrointestinal Adenocarcinoma Incidence and Survival Trends in the Northern Territory, Australia, with Emphasis on Indigenous Peoples. *Cancers* **2022**, *14*, 2870. [CrossRef] [PubMed]
6. Samaan, J.S.; Abboud, Y.; Oh, J.; Jiang, Y.; Watson, R.; Park, K.; Liu, Q.; Atkins, K.; Hendifar, A.; Gong, J.; et al. Pancreatic Cancer Incidence Trends by Race, Ethnicity, Age and Sex in the United States: A Population-Based Study, 2000–2018. *Cancers* **2023**, *15*, 870. [CrossRef] [PubMed]
7. Ten Kate, C.A.; de Klein, A.; de Graaf, B.M.; Doukas, M.; Koivusalo, A.; Pakarinen, M.P.; van der Helm, R.; Brands, T.; Jsselstijn, I.; van Bever, H.Y.; et al. Intrinsic Cellular Susceptibility to Barrett's Esophagus in Adults Born with Esophageal Atresia. *Cancers* **2022**, *14*, 513. [CrossRef] [PubMed]
8. Pocurull, A.; Herrera-Pariente, C.; Carballal, S.; Llach, J.; Sanchez, A.; Carot, L.; Botargues, J.M.; Cuatrecasas, M.; Ocana, T.; Balaguer, F.; et al. Clinical, Molecular and Genetic Characteristics of Early Onset Gastric Cancer: Analysis of a Large Multicenter Study. *Cancers* **2021**, *13*, 3132. [CrossRef] [PubMed]

9. Castle, J.T.; Levy, B.E.; Chauhan, A. Pediatric Neuroendocrine Neoplasms: Rare Malignancies with Incredible Variability. *Cancers* **2022**, *14*, 5049. [CrossRef] [PubMed]
10. Dudzisz-Sledz, M.; Klimczak, A.; Bylina, E.; Rutkowski, P. Treatment of Gastrointestinal Stromal Tumors (Gists): A Focus on Younger Patients. *Cancers* **2022**, *14*, 2831. [CrossRef] [PubMed]
11. Ben-Aharon, I.; van Laarhoven, H.W.M.; Fontana, E.; Obermannova, R.; Nilsson, M.; Lordick, F. Early-Onset Cancer in the Gastrointestinal Tract Is on the Rise-Evidence and Implications. *Cancer Discov.* **2023**, *13*, 538–551. [CrossRef] [PubMed]
12. Barreto, S.G. Young-Onset Pancreatobiliary Cancers—Whereto from Here? *Indian J. Surg.* **2022**, *84*, s349–s351. [CrossRef]
13. Barreto, S.G.; Pandol, S.J. Young-Onset Carcinogenesis-the Potential Impact of Perinatal and Early Life Metabolic Influences on the Epigenome. *Front Oncol.* **2021**, *11*, 653289. [CrossRef]
14. Bergeron, L.A.; Besenbacher, S.; Zheng, J.; Li, P.; Bertelsen, M.F.; Quintard, B.; Hoffman, J.I.; Li, Z.; St Leger, J.; Shao, C.; et al. Evolution of the Germline Mutation Rate across Vertebrates. *Nature* **2023**, *615*, 285–291. [CrossRef]
15. Shirazi, T.N.; Rosinger, A.Y. Reproductive Health Disparities in the USA: Self-Reported Race/Ethnicity Predicts Age of Menarche and Live Birth Ratios, but Not Infertility. *J. Racial Ethn. Health Disparities* **2021**, *8*, 33–46. [CrossRef] [PubMed]

cancers

MDPI

Article

Variation in Treatment Patterns of Patients with Early-Onset Gastric Cancer

Michael LaPelusa [1,*,†], Chan Shen [2,3,†], Erin A. Gillaspie [4], Christopher Cann [5], Eric Lambright [4], A. Bapsi Chakravarthy [6], Michael K. Gibson [5] and Cathy Eng [5]

1 Department of Internal Medicine, Vanderbilt University Medical Center, Nashville, TN 37232, USA
2 Department of Surgery, Penn State College of Medicine, Hershey, PA 17033, USA; cshen@pennstatehealth.psu.edu
3 Department of Public Health Sciences, Penn State College of Medicine, Hershey, PA 17033, USA
4 Division of Thoracic Surgery, Department of Surgery, Vanderbilt University Medical Center, Nashville, TN 37232, USA; erin.a.gillaspie@vumc.org (E.A.G.); eric.lambright@vumc.org (E.L.)
5 Division of Hematology and Oncology, Department of Internal Medicine, Vanderbilt-Ingram Cancer Center, Vanderbilt University Medical Center, Nashville, TN 37232, USA; christopher.g.cann@vumc.org (C.C.); mike.gibson.1@vumc.org (M.K.G.); cathy.eng@vumc.org (C.E.)
6 Department of Radiation Oncology, Vanderbilt-Ingram Cancer Center, Vanderbilt University Medical Center, Nashville, TN 37232, USA; bapsi.chak@vumc.org
* Correspondence: mblapelusa@gmail.com; Tel.: +1-615-322-5000
† These authors contributed equally to this work.

Simple Summary: Gastric cancer is not routinely diagnosed in patients younger than 45. However, the incidence of gastric cancer in young patients is rising. Little is known about the demographic features of young patients diagnosed with gastric cancer. Additionally, the relationship between the therapies these patients receive and their socioeconomic characteristics has not been delineated. We showed that younger patients were more likely to be female, Asian/Pacific Islander, African American, Hispanic, and have advanced-stage disease compared to older patients with gastric cancer. After adjusting for disease stage, we identified differences in receipt of surgery, chemotherapy, and radiation among young patients with gastric cancer based on gender/sex, race/ethnicity, treatment center type, insurance status, and location of residence. Future work should focus on understanding whether these differences were driven by patient choice or alternative reasons.

Citation: LaPelusa, M.; Shen, C.; Gillaspie, E.A.; Cann, C.; Lambright, E.; Chakravarthy, A.B.; Gibson, M.K.; Eng, C. Variation in Treatment Patterns of Patients with Early-Onset Gastric Cancer. *Cancers* **2022**, *14*, 3633. https://doi.org/10.3390/cancers14153633

Academic Editors: Irit Ben Aharon and Savio George Barreto

Received: 23 June 2022
Accepted: 19 July 2022
Published: 26 July 2022

Abstract: Background: Early-onset gastric cancer (EOGC), or gastric cancer in patients younger than 45 years old, is poorly understood and relatively uncommon. Similar to other gastrointestinal malignancies, the incidence of EOGC is rising in Western countries. It is unclear which populations experience a disproportionate burden of EOGC and what factors influence how patients with EOGC are treated. Methods: We conducted a retrospective, population-based study of patients diagnosed with gastric cancer from 2004 to 2018 using the National Cancer Database (NCDB). In addition to identifying unique demographic characteristics of patients with EOGC, we evaluated (using multivariable logistic regression controlling for year of diagnoses, primary site, and stage) how gender/sex, race/ethnicity, treatment facility type, payor status, and location of residence influenced the receipt of surgery, chemotherapy, and radiation. Results: Compared to patients 45–70 and >70 years of age with gastric cancer, patients with EOGC were more likely to be female, Asian/Pacific Islander (PI), African American (AA), Hispanic, uninsured, and present with stage IV disease. On multivariable analysis, several differences among subsets of patients with EOGC were identified. Female patients with EOGC were less likely to receive surgery and chemotherapy than male patients with EOGC. Asian/Pacific Islander patients with EOGC were more likely to receive chemotherapy and less likely to receive radiation than Caucasian patients with EOGC. African American patients were more likely to receive chemotherapy than Caucasian patients with EOGC. Hispanic patients were more likely to receive surgery and chemotherapy and less likely to receive radiation than Caucasian patients with EOGC. Patients with EOGC treated at community cancer centers were more likely to receive surgery and less likely to receive chemotherapy than patients with EOGC treated at academic centers. Uninsured patients with EOGC were more likely to receive surgery and less likely to receive chemotherapy than privately insured patients with EOGC. Patients with EOGC living in locations not adjacent to metropolitan areas were less likely to receive surgery compared to patients with EOGC who resided in metropolitan areas, Conclusions: Patients with EOGC are a demographically distinct population. Treatment of these patients varies significantly based on several demographic factors. Additional analysis is needed to elucidate why particular groups are more affected by EOGC and how treatment decisions are made for, and by, these patients.

Keywords: gastric cancer; early-onset; NCDB; incidence; treatment

1. Introduction

Globally, gastric cancer is a significant public health issue. In 2020, there were over one million new cases and 769,000 deaths, making it the fifth most common cancer and fourth most common cause of cancer-related death [1]. In the United States (US), there are projected to be 26,380 new cases of gastric cancer and 11,090 cancer-related deaths in 2022 [2].

Early-onset gastric cancer (EOGC) is a relatively uncommon phenomenon. One estimate concluded that anywhere between 10–30% of gastric cancer occurs in young patients [3]. However, similarly to several other early-onset gastrointestinal malignancies, the incidence of EOGC is increasing in Western countries [4–9].

Relative to older patients with gastric cancer, less data exist regarding the demographic makeup of young patients with gastric cancer. Additionally, the impact of socioeconomic factors on treatment patterns in this population is unknown. Our primary objective was to delineate how gender/sex, race/ethnicity, treatment center type, insurance status, and residence location contribute to the treatment of patients with EOGC.

2. Methods

2.1. Data Source

This large observational study utilized the National Cancer Database (NCDB) 2004–2018 data set. The NCDB is a national collaboration sponsored by the American Cancer Society and the American College of Surgeons. The NCDB captures approximately 70% of all new cancer diagnoses in the US and is widely accepted as a data source for cancer outcomes research [10]. Data on cancer patients were collected by Commission on Cancer accredited facilities.

2.2. Study Cohort Selection

We identified patients 18 to 90 years old who were diagnosed with gastric cancer from the NCDB 2004–2018 data set. Age cutoffs among population-based analyses of EOGC vary, ranging from 30 to 60 years old [9,11,12]. Given the heterogeneity in definitions of "early-onset", we chose a cutoff of <45 years old. We stratified the sample into three age groups: <45 (EOGC), 45–70 (AOGC), and >70 years of age (LOGC).

2.3. Factors Considered

We considered the following patient demographics and characteristics: age, sex, race/ethnicity (non-Hispanic White hereto referred to as Caucasian, non-Hispanic Black hereto referred to as African American, Hispanic, Asian or Pacific Islander, and unknown), insurance status (uninsured, Medicaid, Medicare, other government, private, unknown), facility type (academic, comprehensive community, and community), location of residence (not metropolitan adjacent, metropolitan adjacent, metropolitan, unknown), and year of diagnosis. We also included the following tumor characteristics: disease stage (stage I, II, III, IV, unknown), tumor location (cardia, non-cardia), and histologic grade (well-differentiated, moderately-differentiated, poorly-differentiated, unknown) in our analysis. We examined the use of surgery, chemotherapy, or radiation individually since data were not available on receipt of bimodal or trimodal therapy.

2.4. Statistical Analysis

We used chi-square tests to examine whether the categorical variables (sex, race/ethnicity, facility type, year of diagnosis, primary payer, location of residence, primary site, stage, use of surgery, chemotherapy and radiation) varied significantly by age groups (EOGC, AOGC, LOGC). When evaluating the effect of specific demographic variables on treatment modality, the multivariable logistic regression models were always adjusted for year of diagnosis, primary site, and stage of cancer. We provide adjusted Odds Ratios (aOR) and 95% Confidence Interval (95% CI). $p \leq 0.05$ was considered statistically significant. Analyses were conducted using SAS software, version 9.4 (SAS Institute, Inc., Cary, CA, USA).

3. Results

233,772 patients were identified from the NCDB between 2004 and 2018. Overall, 114,469 (49%) patients received surgery, 113,053 (48.4%) patients received chemotherapy, and 55,092 (23.6%) patients received radiation therapy.

As displayed in Table 1, females represented a higher proportion of patients with EOGC compared to patients with AOGC and LOGC. A greater percentage of patients with EOGC were Asian/Pacific Islander (PI), African American (AA), and Hispanic relative to patients with AOGC and LOGC. Patients with EOGC demonstrated a higher uninsurance rate than patients with AOGC and LOGC. Patients with EOGC presented with stage IV disease more frequently than patients with AOGC and LOGC.

Table 1. Demographics.

	Age Categories				
	EOGC (*n* = 14,490)	**AOGC (*n* = 118,918)**	**LOGC (*n* = 100,364)**	**Total (*n* = 233,772)**	***p*-Value**
Age at Diagnosis					
Mean (SD)	37.5 (5.80)	60.0 (6.92)	79.4 (5.73)	66.9 (13.63)	
Median	39.0	61.0	79.0	68.0	
Range	18.0, 44.0	45.0, 70.0	71.0, 90.0	18.0, 90.0	
Sex, n (%)					<0.0001
Male	7687 (53.1%)	77,902 (65.5%)	59,243 (59.0%)	144,832 (62.0%)	
Female	6803 (46.9%)	41,016 (34.5%)	41,121 (41.0%)	88,940 (38.0%)	
Race/Ethnicity, n (%)					<0.0001
Hispanic	3600 (24.8%)	12,392 (10.4%)	7107(7.1%)	23,099 (9.9%)	
White non-Hispanic	6351 (43.8%)	73,150 (61.5%)	68,427 (68.2%)	147,928 (63.3%)	
Black non-Hispanic	2398 (16.5%)	18,846 (15.8%)	12,537(12.5%)	33,781 (14.5%)	
Asian/PI non-Hispanic	1326 (9.2%)	7907 (6.6%)	6247 (6.2%)	15,480 (6.6%)	
Unknown	815 (5.6%)	6623 (5.6%)	6046 (6.0%)	13,484 (5.8%)	
Facility Type, n (%)					<0.0001
Community Cancer Program	789 (5.4%)	7404 (6.2%)	7690 (7.7%)	15,883 (6.8%)	
Comprehensive Community Cancer Program	4169 (28.8%)	39,110 (32.9%)	38,085 (37.9%)	81,364 (34.8%)	

Table 1. *Cont.*

	Age Categories				
	EOGC (*n* = 14,490)	**AOGC (*n* = 118,918)**	**LOGC (*n* = 100,364)**	**Total (*n* = 233,772)**	***p*-Value**
Academic/Research Program or Integrated Network Cancer Program	9532 (65.8%)	72,404 (60.9%)	54,589 (54.4%)	136,525 (58.4%)	
Year of Diagnosis, n (%)					<0.0001
2004–2008	4210 (29.1%)	31,212 (26.2%)	30,092 (30.0%)	65,514 (28.0%)	
2009–2013	4783 (33.0%)	40,425 (34.0%)	33,425(33.3%)	78,633 (33.6%)	
2014–2018	5497 (37.9%)	47,281 (39.8%)	36,847 (36.7%)	89,625 (38.3%)	
Primary Payer, n (%)					<0.0001
Not Insured	1695 (11.7%)	6448 (5.4%)	877 (0.9%)	9020 (3.9%)	
Private Insurance	8560 (59.1%)	56,628 (47.6%)	9716 (9.7%)	74,904 (32.0%)	
Medicaid	2961 (20.4%)	12,048 (10.1%)	2369 (2.4%)	17,378 (7.4%)	
Medicare	613 (4.2%)	38,902 (32.7%)	84,847(84.5%)	124,362 (53.2%)	
Other Government	184 (1.3%)	1969 (1.7%)	802 (0.8%)	2955 (1.3%)	
Insurance Status Unknown	477 (3.3%)	2923 (2.5%)	1753 (1.7%)	5153 (2.2%)	
Location of residence, n (%)					<0.0001
Metro counties	12,494 (86.2%)	99,026 (83.3%)	84,832 (84.5%)	196,352 (84.0%)	
Adjacent to metro area	1028 (7.1%)	11,057 (9.3%)	8749 (8.7%)	20,834 (8.9%)	
Not adjacent to metro area	475 (3.3%)	4934 (4.1%)	3928 (3.9%)	9337 (4.0%)	
Unknown	493 (3.4%)	3901(3.3%)	2855 (2.8%)	7249 (3.1%)	
Primary Site, n (%)					<0.0001
Cardia, NOS	3420 (23.6%)	44,808 (37.7%)	31,090 (31.0%)	79,318 (33.9%)	
Non Cardia	11,070 (76.4%)	74,110 (62.3%)	69,274 (69.0%)	154,454 (66.1%)	
Stage, n (%)					<0.0001
Stage I	2341 (16.2%)	24,156 (20.3%)	22,923(22.8%)	49,420 (21.1%)	
Stage II	1450 (10.0%)	15,814 (13.3%)	13,288 (13.2%)	30,552 (13.1%)	
Stage III	2306 (15.9%)	21,550 (18.1%)	14,733 (14.7%)	38,589 (16.5%)	
Stage IV	6229 (43.0%)	40,375 (34.0%)	27,258 (27.2%)	73,862 (31.6%)	
Unknown	2164 (14.9%)	17,023 (14.3%)	22,162 (22.1%)	41,349 (17.7%)	
Surgery, n (%)					<0.0001
No surgery	7149 (49.3%)	54,716 (46.0%)	56,289 (56.1%)	118,154 (50.5%)	
Surgery	7290 (50.3%)	63,596 (53.5%)	43,583 (43.4%)	114,469 (49.0%)	
Unknown	51 (0.4%)	606 (0.5%)	492 (0.5%)	1149 (0.5%)	
Chemotherapy, n (%)					<0.0001
No chemotherapy	4831 (33.3%)	46,103 (38.8%)	62,392 (62.2%)	113,326 (48.5%)	

Table 1. *Cont.*

	Age Categories				
	EOGC (*n* = 14,490)	AOGC (*n* = 118,918)	LOGC (*n* = 100,364)	Total (*n* = 233,772)	*p*-Value
Chemotherapy received	9242 (63.8%)	69,176 (58.2%)	34,635(34.5%)	113,053 (48.4%)	
Unknown	417 (2.9%)	3639 (3.1%)	3337 (3.3%)	7393 (3.2%)	
Radiation, n (%)					<0.0001
No radiation	10,974 (75.7%)	83,467 (70.2%)	77,662 (77.4%)	172,103 (73.6%)	
Radiation received	3127 (21.6%)	32,073 (27.0%)	19,892 (19.8%)	55,092 (23.6%)	
Unknown	389 (2.7%)	3378 (2.8%)	2810 (2.8%)	6577 (2.8%)	

As displayed in Table 2, female patients with EOGC were less likely to receive surgery and chemotherapy but more likely to receive radiation compared to male patients with EOGC. Compared to Caucasian patients, Asian/PI patients with EOGC were more likely to receive chemotherapy and less likely to receive radiation, AA patients with EOGC were more likely to receive chemotherapy, and Hispanic patients with EOGC were more likely to receive surgery and chemotherapy and less likely to receive radiation. Patients with EOGC treated at community cancer centers were more likely to receive surgery and less likely to receive chemotherapy than patients with EOGC treated at academic centers. Patients with EOGC treated at comprehensive community centers were more likely to receive surgery and less likely to receive radiation than patients with EOGC treated at academic centers. Compared to privately insured patients with EOGC, uninsured patients with EOGC were more likely to receive surgery and less likely to receive chemotherapy. Patients with EOGC who had Medicaid were more likely to receive surgery than privately insured patients. Patients with EOGC who resided in locations not adjacent to metropolitan areas were less likely to receive surgery than patients living in metropolitan areas.

Table 2. Treatment patterns among patients with EOGC.

Variable	Categories	Odds Ratio; 95% CI; *p*-Value		
		Surgery	Chemotherapy	Radiation
Age	(Continuous)	1.00; [0.99, 1.00]; 0.334	1.00; [0.99, 1.00]; 0.441	0.99; [0.98, 0.99]; 0.033
Gender/Sex	Female	0.89; [0.81, 0.97]; 0.008	0.80; [0.74, 0.87]; <0.001	1.41; [1.29, 1.56]; <0.001
	Male		Reference	
Race/Ethnicity	Asian/PI	1.08; [0.93, 1.26]; 0.321	1.66; [1.43, 1.92]; <0.001	0.74; [0.63, 0.87]; <0.001
	African American	1.07; [0.95, 1.21]; 0.274	1.21; [1.08, 1.36]; <0.001	0.94; [0.82, 1.08]; 0.385
	Hispanic	1.42; [1.26, 1.59]; <0.001	1.52; [1.37, 1.69]; <0.001	0.82; [0.73, 0.93]; 0.002
	Non-Hispanic White		Reference	
Facility Type	Community	1.24; [1.02, 1.50]; 0.029	0.80; [0.67, 0.95]; 0.013	0.84; [0.69, 1.03]; 0.100
	Comprehensive Community	1.15; [1.05, 1.27]; 0.003	1.02; [0.93, 1.12]; 0.662	0.78; [0.71, 0.86]; <0.001
	Academic		Reference	
Payor Status	Uninsured	1.92; [1.67, 2.22]; <0.001	0.78; [0.68, 0.88]; <0.001	1.07; [0.92, 1.25]; 0.132
	Medicaid	1.69; [1.51, 1.89]; <0.001	0.90; [0.82, 1.00]; 0.061	1.01; [0.89, 1.13]; 0.906
	Medicare	1.44; [1.17, 1.78]; <0.001	0.50; [0.41, 0.60]; <0.001	1.08; [0.86, 1.36]; 0.497

Table 2. *Cont.*

Variable	Categories	Odds Ratio; 95% CI; *p*-Value		
		Surgery	Chemotherapy	Radiation
	Other Government	1.61; [1.10, 2.35]; 0.014	0.94; [0.66, 1.36]; 0.759	0.58; [0.40, 0.85]; 0.005
	Unknown	1.93; [1.49, 2.49]; <0.001	0.85; [0.67, 1.09]; 0.209	1.16; [0.88, 1.53]; 0.282
	Private		Reference	
Location	Not Metro Adjacent	0.69; [0.54, 0.89]; 0.004	1.03; [0.82, 1.30]; 0.781	1.00; [0.78, 1.29]; 0.974
	Metro Adjacent	0.93; [0.79, 1.10]; 0.383	1.06; [0.91, 1.24]; 0.448	0.94; [0.79, 1.11]; 0.442
	Metro		Reference	

Selected results are presented in this table. The multivariable logistic regression also controlled for year of diagnosis, primary site, and stage of cancer.

4. Discussion

4.1. Demographic Characteristics

Patients with EOGC display unique clinical features. We found that patients with EOGC were more likely to be female, Asian/PI, AA, Hispanic, uninsured, and present with stage IV disease versus patients with AOGC and LOGC. Our analysis was consistent with others that showed EOGC is more common in females, more likely to be diagnosed at an advanced stage and have a disproportionate effect on uninsured patients, African Americans, and Hispanic patients [13–16]. Others have shown that EOGC displays unique genomic features. For example, tumors of patients with EOGC are more likely to have a diffuse histologic subtype and include signet ring cells, more likely to contain mutated *CDH1*, *BANP*, *MUC5B*, and *TGFBR1* genes, and less likely to contain microsatellite instability [9,17–19]. While smoking and alcohol use are known modifiable risk factors for the development of gastric cancer, particularly in the US, where the prevalence of *Helicobacter pylori* infection is relatively low, modifiable risk factors such as smoking and alcohol use were not found to be associated with the development of EOGC in an analysis of the Behavioral Risk Factor Surveillance System [9]. Some have speculated that EOGC is associated with proton pump inhibitor use via increased gastrin production and subsequent gastrin-induced carcinogenesis. However, conflicting data exist on this topic [20,21]. Others have purported there to be an association between Epstein Barr Virus and EOGC—however, these data are not consistent which may be secondary to variability between tumor samples in the Cancer Genome Atlas, Hong Kong Cancer Registry, and Asian Cancer Research Group cohorts [9,22–24]. Limited data exist on how patients with EOGC are treated compared to older patients. One previous analysis of SEER data showed that patients with EOGC who underwent surgery received more adjuvant radiation compared to older patients with gastric cancer [25]. Another analysis in China showed that patients with EOGC were more likely to receive chemotherapy than older patients, a finding possibly related to better performance status in younger patients [26].

4.2. Treatment by Gender/Sex

We found female patients with EOGC were less likely to receive surgery and chemotherapy but more likely to receive radiation than males with EOGC [Table 2]. Several epidemiological studies of gastric cancer treatment patterns have similarly identified an association between the receipt of less surgery and chemotherapy with female gender/sex. In an NCDB analysis of patients with stage Ib-III gastric cancer of all ages, female patients were less likely to receive perioperative chemotherapy than males [27]. Female patients of all ages that underwent surgery with curative intent in the Netherlands were also less likely than males to receive perioperative chemotherapy and were more likely to undergo partial gastrectomy (rather than total gastrectomy). However, these differences were not statistically significant after adjusting for clinicopathologic factors such as clinical stage [28].

In another Dutch study of treatment allocation, female patients with unresectable gastric cancer were less likely to receive chemotherapy compared to males [29].

4.3. Treatment by Race/Ethnicity

We found Asian/PI patients with EOGC and AOGC were more likely to receive chemotherapy than Caucasian patients with EOGC and AOGC, respectively (Tables 2 and S1), and Asian/PI patients with EOGC, AOGC, and LOGC were less likely to receive radiation compared to Caucasian patients with EOGC, AOGC, and LOGC, respectively (Tables 2, S1 and S2). AA patients with EOGC and AOGC were more likely than Caucasian patients with EOGC and AOGC to receive chemotherapy, respectively (Tables 2 and S1). Hispanic patients with EOGC and AOGC were more likely to receive surgery and chemotherapy than Caucasian patients with EOGC and AOGC, respectively (Tables 2 and S1). Previous analyses of treatment differences of gastric cancer by race/ethnicity are not stratified by age. With this limitation, others have consistently found that Asian/PI patients with gastric cancer are more likely to receive therapy than other groups [30–33]. In the aforementioned NCDB analysis of patients of all ages with stage Ib-III gastric cancer undergoing surgery, Asian/PI and AA patients were less likely than Caucasian patients to receive perioperative chemotherapy while no difference was found among Hispanic patients [27]. It is known that Asian American, African American, and Hispanic patients receive hospice and palliative care at lower rates compared to Caucasian patients which some have theorized is related to differences in knowledge, cultural beliefs, and treatment preferences [34,35]. Assuming the utilization of hospice and palliative care is a surrogate for the receipt of less treatment, it is possible that this disparity in hospice and palliative care utilization is an explanation for our findings (regarding increased receipt of treatment among patients who are Asian/PI, African American, and Hispanic compared to Caucasian patients). Communication barriers and assumptions made by patients and their oncologists likely also play a role in the differences we observed.

4.4. Treatment by Center Type

We found patients with EOGC, AOGC, and LOGC treated at community cancer centers were more likely to receive surgery and less likely to receive chemotherapy than patients with EOGC, AOGC, and LOGC, treated at academic centers, respectively (Tables 2, S1 and S2). In England, patients diagnosed with esophageal and gastric cancers in non-academic hospitals did not have a lower chance of having surgery than those diagnosed in an academic hospital [36]. Several studies in the Netherlands have identified patterns in the treatment of gastric cancer by hospital type and found that patients with gastric cancer treated at high-volume hospitals were more likely to receive systemic therapy and surgery compared to hospitals with lower volume [37,38]. Academic centers are more likely to have enroll patients on clinical trials and offer treatment options that are not available in community cancer centers, which may help explain our findings.

4.5. Treatment by Payor Status

We found patients with EOGC, AOGC, and LOGC who were uninsured or had Medicaid were more likely to receive surgery and less likely to receive chemotherapy than insured patients with EOGC, AOGC, and LOGC, respectively (Tables 2, S1 and S2). In the Netherlands, younger age and higher socioeconomic status (SES) were independent factors for receiving treatment in patients with esophageal and gastric cancer [38,39]. Notably, patients with gastric cancer who lack insurance have been shown to have worse survival outcomes and receive less therapy compared to insured patients [40,41]. Insurance status plays a role in the type of treatment patients can receive (as well as where they can receive it).

4.6. Treatment by Location

We found patients with EOGC and LOGC residing in locations that were not adjacent to metropolitan areas were less likely to receive surgery than those residing in metropolitan areas (Tables 2 and S2). In analyses of SEER and California Cancer Registry data, patients of all ages with gastric cancer residing in rural areas were also less likely to receive surgery compared to those in urban areas, which was attributed to lower levels of educational attainment, lower median household income, longer commute times, less contact with oncology providers, and less access to health insurance [42,43].

5. Conclusions

Our study represents the most comprehensive to date regarding the unique treatment patterns of patients with EOGC. As an entity, EOGC displays many alarming features—the incidence of this entity is increasing, these patients tend to present with late-stage disease, and risk factors are not well defined.

Our study had several important limitations. Most notably, individual-level data regarding the treatment sequence for each patient are not available in the NCDB, nor are data regarding environmental risk factors and tumor genomic information.

We found dramatic, statistically significant differences regarding how patients with EOGC are treated after adjusting for stage, tumor location, and year of diagnosis. However, the reasons why subgroups of patients with EOGC were treated differently is unclear. Ultimately, the complex interplay between intrinsic patient perceptions of treatment combined with external forces such as residence in a resource-limited setting, inadequate health insurance, and bias on the part of providers are likely intertwined. More research to untangle this complex narrative is warranted to characterize which factors play a role in the pursuit and receipt of treatment from both the patient and oncologist perspectives. Additionally, developing effective cultural awareness, minimizing assumptions, and recognizing differences in communication preferences are important to mitigate discrimination against, and implicit bias towards, marginalized patient populations. Investing in educational programs and healthcare systems to ensure patients have every opportunity to access high-quality care, as well as clinical trials, is imperative.

Supplementary Materials: The following supporting information can be downloaded at: https://www.mdpi.com/article/10.3390/cancers14153633/s1, Table S1: Treatment Patterns Among Patients with AOGC; Table S2: Treatment Patterns Among Patients with LOGC.

Author Contributions: M.L., C.S. and C.E. were involved with study conception and design. M.L., C.S., E.A.G., C.C. and A.B.C. were involved with draft manuscript preparation. E.L., M.K.G. and C.E. were involved with editing. All authors were involved with interpreting results. All authors have read and agreed to the published version of the manuscript.

Funding: This research received no external funding.

Institutional Review Board Statement: Ethical review and approval were waived for this study due to the utilization of a publicly available, de-identified database.

Informed Consent Statement: Not applicable.

Data Availability Statement: A publicly available dataset was analyzed in this study. This data can be found here: https://www.facs.org/quality-programs/cancer-programs/national-cancer-database/publicaccess/, accessed on 3 June 2022.

Conflicts of Interest: The authors have no conflict to disclose in relation to this study.

References

1. Sung, H.; Ferlay, J.; Siegel, R.L.; Laversanne, M.; Soerjomataram, I.; Jemal, A.; Bray, F. Global cancer statistics 2020: GLOBOCAN estimates of incidence and mortality worldwide for 36 cancers in 185 countries. *CA Cancer J. Clin.* **2021**, *71*, 209–249. [CrossRef]
2. Siegel, R.L.; Miller, K.D.; Fuchs, H.E.; Jemal, A. Cancer statistics, 2022. *CA Cancer J. Clin.* **2022**, *72*, 7–33. [CrossRef]
3. Kokkola, A.; Sipponen, P. Gastric carcinoma in young adults. *Hepato-Gastroenterology* **2001**, *48*, 1552–1555.

4. Arnold, M.; Park, J.Y.; Camargo, M.C.; Lunet, N.; Forman, D.; Soerjomataram, I. Is gastric cancer becoming a rare disease? A global assessment of predicted incidence trends to 2035. *Gut* **2020**, *69*, 823–829. [CrossRef]
5. Heer, E.V.; Harper, A.S.; Sung, H.; Jemal, A.; Fidler-Benaoudia, M.M. Emerging cancer incidence trends in Canada: The growing burden of young adult cancers. *Cancer* **2020**, *126*, 4553–4562. [CrossRef]
6. LaPelusa, M.; Shen, C.; Arhin, N.D.; Cardin, D.; Tan, M.; Idrees, K.; Geevarghese, S.; Chakravarthy, B.; Berlin, J.; Eng, C. Trends in the Incidence and Treatment of Early-Onset Pancreatic Cancer. *Cancers* **2022**, *14*, 283. [CrossRef]
7. Codipilly, D.C.; Sawas, T.; Dhaliwal, L.; Johnson, M.L.; Lansing, R.; Wang, K.K.; Leggett, C.L.; Katzka, D.A.; Iyer, P.G. Epidemiology and outcomes of young-onset esophageal adenocarcinoma: An analysis from a population-based database. *Cancer Epidemiol. Prev. Biomark.* **2021**, *30*, 142–149. [CrossRef]
8. Siegel, R.L.; Fedewa, S.A.; Anderson, W.F.; Miller, K.D.; Ma, J.; Rosenberg, P.S.; Jemal, A. Colorectal cancer incidence patterns in the United States, 1974–2013. *JNCI J. Natl. Cancer Inst.* **2017**, *109*, djw322. [CrossRef]
9. Bergquist, J.R.; Leiting, J.L.; Habermann, E.B.; Cleary, S.P.; Kendrick, M.L.; Smoot, R.L.; Nagorney, D.M.; Truty, M.J.; Grotz, T.E. Early-onset gastric cancer is a distinct disease with worrisome trends and oncogenic features. *Surgery* **2019**, *166*, 547–555. [CrossRef]
10. Boffa, D.J.; Rosen, J.E.; Mallin, K.; Loomis, A.; Gay, G.; Palis, B.; Thoburn, K.; Gress, D.; McKellar, D.P.; Shulman, L.N. Using the National Cancer Database for outcomes research: A review. *JAMA Oncol.* **2017**, *3*, 1722–1728. [CrossRef]
11. Theuer, C.P.; de Virgilio, C.; Keese, G.; French, S.; Arnell, T.; Tolmos, J.; Klein, S.; Powers, W.; Oh, T.; Stabile, B.E. Gastric adenocarcinoma in patients 40 years of age or younger. *Am. J. Surg.* **1996**, *172*, 473–477. [CrossRef]
12. Rona, K.A.; Schwameis, K.; Zehetner, J.; Samakar, K.; Green, K.; Samaan, J.; Sandhu, K.; Bildzukewicz, N.; Katkhouda, N.; Lipham, J.C. Gastric cancer in the young: An advanced disease with poor prognostic features. *J. Surg. Oncol.* **2017**, *115*, 371–375. [CrossRef]
13. Torrejon, N.V.; Deshpande, S.; Wei, W.; Tullio, K.; Kamath, S.D. Proportion of early-onset gastric and esophagus cancers has changed over time with disproportionate impact on Black and Hispanic patients. *JCO Oncol. Pract.* **2022**, *18*, e759–e769. [CrossRef]
14. Holowatyj, A.N.; Ulrich, C.M.; Lewis, M.A. Racial/ethnic patterns of young-onset noncardia gastric cancer. *Cancer Prev. Res.* **2019**, *12*, 771–780. [CrossRef]
15. Giryes, A.; Oweira, H.; Mannhart, M.; Decker, M.; Abdel-Rahman, O. Exploring the differences between early-onset gastric cancer and traditional-onset gastric cancer. *J. Gastrointest. Oncol.* **2018**, *9*, 1157. [CrossRef]
16. De, B.; Rhome, R.; Jairam, V.; Özbek, U.; Holcombe, R.F.; Buckstein, M.; Ang, C. Gastric adenocarcinoma in young adult patients: Patterns of care and survival in the United States. *Gastric Cancer* **2018**, *21*, 889–899. [CrossRef]
17. Cho, S.Y.; Park, J.W.; Liu, Y.; Park, Y.S.; Kim, J.H.; Yang, H.; Um, H.; Ko, W.R.; Lee, B.I.; Kwon, S.Y. Sporadic early-onset diffuse gastric cancers have high frequency of somatic CDH1 alterations, but low frequency of somatic RHOA mutations compared with late-onset cancers. *Gastroenterology* **2017**, *153*, 536–549.e526. [CrossRef]
18. Huang, Q.; Zheng, X.; Jiao, Y.; Lei, Y.; Li, X.; Bi, F.; Guo, F.; Wang, G.; Liu, M. A Distinct Clinicopathological Feature and Prognosis of Young Gastric Cancer Patients Aged $\leq$ 45 Years Old. *Front. Oncol.* **2021**, *11*, 674224. [CrossRef]
19. Ma, Z.; Liu, X.; Paul, M.E.; Chen, M.; Zheng, P.; Chen, H. Comparative investigation of early-onset gastric cancer. *Oncol. Lett.* **2021**, *21*, 374. [CrossRef]
20. Waldum, H.L. The increase in early-onset gastric carcinomas from 1995 is probably due to the introduction of proton pump inhibitors. *Surgery* **2020**, *168*, 568–569. [CrossRef]
21. MacArthur, T.A.; Harmsen, W.S.; Mandrekar, J.; Abraha, F.; Grotz, T.E. Association of Common Medications and the Risk of Early-Onset Gastric Cancer: A Population-Based Matched Study. *J. Cancer Epidemiol.* **2021**, *2021*, 2670502. [CrossRef] [PubMed]
22. Wang, K.; Yuen, S.T.; Xu, J.; Lee, S.P.; Yan, H.H.; Shi, S.T.; Siu, H.C.; Deng, S.; Chu, K.M.; Law, S. Whole-genome sequencing and comprehensive molecular profiling identify new driver mutations in gastric cancer. *Nat. Genet.* **2014**, *46*, 573–582. [CrossRef] [PubMed]
23. Cristescu, R.; Lee, J.; Nebozhyn, M.; Kim, K.-M.; Ting, J.C.; Wong, S.S.; Liu, J.; Yue, Y.G.; Wang, J.; Yu, K. Molecular analysis of gastric cancer identifies subtypes associated with distinct clinical outcomes. *Nat. Med.* **2015**, *21*, 449–456. [CrossRef]
24. Zhao, L.-Y.; Hu, J.-K. Early-onset gastric cancer is quite distinct from late-onset gastric cancer. *Surgery* **2020**, *167*, 883. [CrossRef] [PubMed]
25. Al-Refaie, W.B.; Hu, C.-Y.; Pisters, P.W.; Chang, G.J. Gastric adenocarcinoma in young patients: A population-based appraisal. *Ann. Surg. Oncol.* **2011**, *18*, 2800–2807. [CrossRef]
26. Jiang, Y.; Xie, J.; Huang, W.; Chen, H.; Xi, S.; Li, T.; Chen, C.; Sun, Z.; Hu, Y.; Liu, W. Chemotherapy Use and Survival Among Young and Middle-Aged Patients With Gastric Cancer. *Clin. Transl. Gastroenterol.* **2020**, *11*, e00253. [CrossRef]
27. Sherman, K.L.; Merkow, R.P.; Bilimoria, K.Y.; Wang, C.E.; Mulcahy, M.F.; Benson, A.B.; Bentrem, D.J. Treatment trends and predictors of adjuvant and neoadjuvant therapy for gastric adenocarcinoma in the United States. *Ann. Surg. Oncol.* **2013**, *20*, 362–370. [CrossRef]
28. Kalff, M.C.; Wagner, A.D.; Verhoeven, R.H.; Lemmens, V.E.; van Laarhoven, H.W.; Gisbertz, S.S.; van Berge Henegouwen, M.I. Sex differences in tumor characteristics, treatment, and outcomes of gastric and esophageal cancer surgery: Nationwide cohort data from the Dutch Upper GI Cancer Audit. *Gastric Cancer* **2021**, *25*, 22–32. [CrossRef]
29. Dijksterhuis, W.P.M.; Kalff, M.C.; Wagner, A.D.; Verhoeven, R.H.A.; Lemmens, V.E.P.P.; van Oijen, M.G.H.; Gisbertz, S.S.; van Berge Henegouwen, M.I.; van Laarhoven, H.W.M. Gender Differences in Treatment Allocation and Survival of Advanced Gastroesophageal Cancer: A Population-Based Study. *JNCI J. Natl. Cancer Inst.* **2021**, *113*, 1551–1560. [CrossRef]

30. Gill, S.; Shah, A.; Le, N.; Cook, E.F.; Yoshida, E.M. Asian ethnicity–related differences in gastric cancer presentation and outcome among patients treated at a Canadian Cancer Center. *J. Clin. Oncol.* **2003**, *21*, 2070–2076. [CrossRef]
31. Kim, J.; Sun, C.-L.; Mailey, B.; Prendergast, C.; Artinyan, A.; Bhatia, S.; Pigazzi, A.; Ellenhorn, J. Race and ethnicity correlate with survival in patients with gastric adenocarcinoma. *Ann. Oncol.* **2010**, *21*, 152–160. [CrossRef] [PubMed]
32. Klapheke, A.K.; Carvajal-Carmona, L.G.; Cress, R.D. Racial/ethnic differences in survival among gastric cancer patients in california. *Cancer Causes Control* **2019**, *30*, 687–696. [CrossRef] [PubMed]
33. Zhang, G.; Zhao, X.; Li, J.; Yuan, Y.; Wen, M.; Hao, X.; Li, P.; Zhang, A. Racial disparities in stage-specific gastric cancer: Analysis of results from the Surveillance Epidemiology and End Results (SEER) program database. *J. Investig. Med.* **2017**, *65*, 991–998. [CrossRef] [PubMed]
34. Johnson, K.S. Racial and ethnic disparities in palliative care. *J. Palliat. Med.* **2013**, *16*, 1329–1334. [CrossRef] [PubMed]
35. Medicare Payment Advisory Commission (US). *Report to the Congress, Medicare Payment Policy*; Medicare Payment Advisory Commission: Washington, DC, USA, 2003.
36. Bachmann, M.; Alderson, D.; Edwards, D.; Wotton, S.; Bedford, C.; Peters, T.; Harvey, I. Cohort study in South and West England of the influence of specialization on the management and outcome of patients with oesophageal and gastric cancers. *J. Br. Surg.* **2002**, *89*, 914–922. [CrossRef] [PubMed]
37. Dijksterhuis, W.P.; Verhoeven, R.H.; Pape, M.; Slingerland, M.; Mohammad, N.H.; de Vos-Geelen, J.; Beerepoot, L.V.; van Voorthuizen, T.; Creemers, G.-J.; Lemmens, V.E. Hospital volume and beyond first-line palliative systemic treatment in metastatic oesophagogastric adenocarcinoma: A population-based study. *Eur. J. Cancer* **2020**, *139*, 107–118. [CrossRef]
38. Bus, P.; Aarts, M.J.; Lemmens, V.E.; van Oijen, M.G.; Creemers, G.-J.; Nieuwenhuijzen, G.A.; van Baal, J.W.; Siersema, P.D. The effect of socioeconomic status on staging and treatment decisions in esophageal cancer. *J. Clin. Gastroenterol.* **2012**, *46*, 833–839. [CrossRef]
39. Trip, A.K.; Stiekema, J.; Visser, O.; Dikken, J.L.; Cats, A.; Boot, H.; Van Sandick, J.W.; Jansen, E.P.; Verheij, M. Recent trends and predictors of multimodality treatment for oesophageal, oesophagogastric junction, and gastric cancer: A Dutch cohort-study. *Acta Oncol.* **2015**, *54*, 1754–1762. [CrossRef]
40. Findakly, D.; Forlemu, A.; Kosa, D.; Amar, S. *The Impact of Sociodemographic Characteristics on Gastric Cancer Outcomes: A Retrospective Review of a Large Single Center Study*; American Society of Clinical Oncology: Alexandria, VA, USA, 2021.
41. Parikh-Patel, A.; Morris, C.R.; Kizer, K.W. Disparities in quality of cancer care: The role of health insurance and population demographics. *Medicine* **2017**, *96*, e9125. [CrossRef]
42. Minhas, A.A.; Fatima, Z.; Kommineni, S.K.; Ahmad, Z.; Minhas, S.A. The Association of Rural-Urban Inhabitation With Gastric Adenocarcinoma Mortality and Treatment: A Surveillance, Epidemiology, and End Results (SEER)-Based Study. *Cureus* **2021**, *13*, e18571. [CrossRef]
43. Parikh-Patel, A.; Morris, C.R.; Kizer, K.W.; Wun, T.; Keegan, T.H. Urban–Rural Variations in Quality of Care Among Patients With Cancer in California. *Am. J. Prev. Med.* **2021**, *61*, e279–e288. [CrossRef] [PubMed]

cancers

MDPI

Article

Pancreatic Cancer Incidence Trends by Race, Ethnicity, Age and Sex in the United States: A Population-Based Study, 2000–2018

Jamil S. Samaan [1], Yazan Abboud [1], Janice Oh [1], Yi Jiang [1], Rabindra Watson [1], Kenneth Park [1], Quin Liu [1], Katelyn Atkins [2], Andrew Hendifar [3], Jun Gong [3], Arsen Osipov [3], Debiao Li [4], Nicholas N. Nissen [5], Stephen J. Pandol [1], Simon K. Lo [1] and Srinivas Gaddam [1,*]

1 Karsh Division of Gastroenterology and Hepatology, Cedars-Sinai Medical Center, 8700 Beverly Blvd, Suite 7705, Los Angeles, CA 90048, USA
2 Department of Radiation Oncology, Cedars-Sinai Medical Center, Los Angeles, CA 90048, USA
3 Department of Medicine, Samuel Oschin Comprehensive Cancer Institute, Cedars-Sinai Medical Center, Los Angeles, CA 90048, USA
4 Biomedical Imaging Research Institute, Cedars-Sinai Medical Center, Los Angeles, CA 90048, USA
5 Department of Hepatobiliary and Pancreatic Surgery, Cedars-Sinai Medical Center, Los Angeles, CA 90048, USA
* Correspondence: srinivas.gaddam@cshs.org

Simple Summary: Pancreatic cancer (PC) incidence is increasing at a greater rate in young women compared to young men. We aimed to understand the association of race and ethnicity with these trends by performing race, ethnicity and age-specific analysis using the SEER 21 database. We organized race and ethnicity groups by Non-Hispanic White (White), Non-Hispanic Black (Black) and Hispanic, and age groups as older adults (age $\geq$ 55 years) and younger adults (age < 55 years). We found a greater rate of increase in PC incidence among young women compared to young men among all race and ethnicity groups, although young Hispanic and Black women experienced a disproportionately greater increase. When comparing trends among women from all race and ethnicity groups, young Hispanic women experienced a greater rate of increase in PC incidence compared to young White and Black women. Our study demonstrates the association of race and ethnicity with PC incidence trends and highlights the disproportionate burden of disease on young women of color.

Abstract: Background and aims: Pancreatic cancer (PC) incidence is increasing at a greater rate in young women compared to young men. We performed a race- and ethnicity-specific evaluation of incidence trends in subgroups stratified by age and sex to investigate the association of race and ethnicity with these trends. Methods: Age-adjusted PC incidence rates (IR) from the years 2000 to 2018 were obtained from the SEER 21 database. Non-Hispanic White (White), Non-Hispanic Black (Black) and Hispanic patients were included. Age categories included older (ages $\geq$ 55) and younger (ages < 55) adults. Time-trends were described as annual percentage change (APC) and average APC (AAPC). Results: Younger White [AAPC difference = 0.73, p = 0.01)], Black [AAPC difference = 1.96, p = 0.01)] and Hispanic [AAPC difference = 1.55, p = 0.011)] women experienced a greater rate of increase in IR compared to their counterpart men. Younger Hispanic women experienced a greater rate of increase in IR compared to younger Black women [AAPC difference = −1.28, p = 0.028)] and younger White women [AAPC difference = −1.35, p = 0.011)]. Conclusion: Younger women of all races and ethnicities experienced a greater rate of increase in PC IR compared to their counterpart men; however, younger Hispanic and Black women experienced a disproportionately greater increase. Hispanic women experienced a greater rate of increase in IR compared to younger Black and White women.

Keywords: cancer disparity; sex disparity; gender; pancreatic cancer trends; ethnicity; race/racial; white/caucasian; black/african american; hispanic/latinx; disparity/disparities

Citation: Samaan, J.S.; Abboud, Y.; Oh, J.; Jiang, Y.; Watson, R.; Park, K.; Liu, Q.; Atkins, K.; Hendifar, A.; Gong, J.; et al. Pancreatic Cancer Incidence Trends by Race, Ethnicity, Age and Sex in the United States: A Population-Based Study, 2000–2018. *Cancers* **2023**, *15*, 870. https://doi.org/10.3390/cancers15030870

Academic Editors: Shailesh Singh and Rajesh Singh

Received: 20 December 2022
Revised: 22 January 2023
Accepted: 27 January 2023
Published: 31 January 2023

1. Introduction

Despite substantial diagnostic and therapeutic medical advancements in recent decades, pancreatic cancer (PC) remains a deadly disease with an estimated 5-year survival rate of 10.8% [1]. Patients are often diagnosed after regional or distant metastasis, with only 11.3% diagnosed at an early stage [1]. Patients with PC accrue more than 15 times the healthcare costs of patients without PC [2]. Furthermore, these patients suffer decreases in quality in all domains of life and worse psychological quality of life when compared to patients with other cancers [3]. In the year 2021, PC will be the 11th and 3rd overall leading cause of cancer incidence and mortality in the United States (US), respectively [1].

Our recent study examining national time trends in PC incidence using the Surveillance, Epidemiology and End Result (SEER) database revealed sex disparity in PC incidence trends over the past two decades [4]. When stratified by age and sex, younger women (ages < 55) were found to have a significantly higher relative rate of increase in PC incidence compared to younger men. Interestingly, this sex disparity trend was not observed in the older population (ages ≥ 55), posing the possibility of a distinct driver(s) in incidence that may be unique to younger women. However, this study does not give information on whether racial differences accounted for the sharp rise in incidence in younger women.

Racial disparities in the incidence and outcomes of multiple gastrointestinal malignancies have been previously demonstrated, including higher PC incidence and mortality rates among Black compared to White persons [5,6]. Therefore, the aim of the current study was to perform race- and ethnicity-specific evaluations of incidence trends in subgroups stratified by age and sex using a large population-based SEER database.

2. Materials and Methods

2.1. Data Source

The SEER program is a cancer incidence and survival database compiled from population-based cancer registries across the US and provided by the National Cancer Institute (NCI). We used the SEER 21, which covers 36.4% of the US population and reports cancer cases from the year 2000 to 2018 [7,8]. The SEER 21 database is the most comprehensive compared to the remaining available databases, providing data from more geographical regions and including a larger proportion of the US population. Further, this database provides information on reporting delays. Given this data is publicly available and is de-identified, according to Cedars-Sinai institutional policy, this study was considered exempt from the institutional review board's (IRB) full or expedited review protocol. This is in accordance with the recommendations of the National Human Research Protections Advisory Committee.

2.2. Definitions

Incidence rates are reported as the number of cases per 100,000 population. The percent change in PC incidence between two consecutive years was defined as the annual percent change (APC), which describes magnitude as well as directionality. The average APC (AAPC) was calculated using the overall change in incidence divided by the total number of years in the study period. We organized race and ethnicity groups into the following: Non-Hispanic White (White), Non-Hispanic Black (Black) and Hispanic, which includes White Hispanic and Black Hispanic patients [9]. Due to the small sample size, the patient groups Asian/Pacific Islander, American Indian/Alaskan Native and Native Hawaiian were not included in our analysis. Age stratification was performed by using an age cut-off of 55 years: Older adults (age ≥ 55 years) and younger adults (age < 55 years). This cutoff was determined based on precedent from previous publications as well as the examination of the distribution of cases by age groups as shown in Figure S1 [4,10].

2.3. Data Retrieval and Study Period

PC incidence data from 1 January 2000–31 December 2018 were retrieved using the software SEER*Stat (NCI), version 8.3.9.2. PC cases were identified using the International Classification of Diseases for Oncology, third edition, "Site Recode ICD-O-3/WHO 2008 classification".

2.4. Statistical Analysis

The number of cancer cases is presented as frequencies and percentages, while incidence rates are presented as cases per 100,000 population. The software SEER*Stat (NCI), version 8.3.9.2, was used to calculate the annual PC incidence. Incidence rates were age-adjusted for the 2000 US population and for reporting delays. Joinpoint Regression Program v4.9.0.0 was used to generate best-fit models on a logarithmic scale [11]. Parametric estimations were used to calculate APC and AAPC for the study period [12,13]. The statistical difference from zero was assessed using a two-sided t-test for the APC and AAPC, while the tests of parallelism and coincidence were used to assess if trends were parallel or identical, respectively, between the AAPCs of segmented-linear trends [14]. The test of parallelism determines if two trends are parallel to each other by evaluating if the trends have a statistically significant difference in slope. The test of coincidence, sometimes referred to as the test of identicalness, assesses if the rates of one trend are identical or statistically different from the rates of the other trend. The two p-values generated from these two tests are comparative in nature between trends in subgroups. The concepts of parallelism and coincidence are illustrated in Figure S2.

The test of parallelism was conducted using log-linear models on the log-transformed scale of the APCs. Subsequently, the Joinpoint Regression Program back-transforms the results to the original scale and provides a report [14,15]. To estimate the statistical significance of the difference between AAPCs, Taylor Series expansion was used [16]. These methods were used to examine sex disparity in PC incidence trends when stratified by age, race and ethnicity, and further analysis examined race and ethnic disparity in cancer incidence among women stratified by age, race and ethnicity. A 2-sided p-value of less than 0.05 was considered statistically significant.

3. Results

3.1. Demographics

A total of 283,817 PC cases were reported during the study period. The majority occurred in women (50.1%) and White patients (72.7%). A total of 88.6% of patients were diagnosed at an age greater than or equal to 55, with a median age of 71 years old. The SEER 21 database reports cancer data on 36.4% of the US population, including 33.6%, 44.7%, and 46.7% of the White, Black and Hispanic populations in the US, respectively [8]. A summary of PC incidence rates stratified by age, sex, race and ethnicity is shown in Table 1.

Table 1. Summary of age-adjusted pancreatic cancer incidence rates per 100,000 people stratified by age, sex, race and ethnicity.

	All Ages	Age ≥ 55	Age < 55
	All Races and Ethnicities		
Women			
Cases (%)	142,297 (50.1)	128,204 (45.2)	14,035 (4.9)
Incidence Rate (95% CI)	11.22 (11.16, 11.28)	46.49 (46.23, 46.75)	2.25 (2.21, 2.28)
Men			
Cases (%)	141,520 (49.9)	123,156 (43.4)	18,334 (6.5)
Incidence Rate (95% CI)	14.32 (14.24, 14.39)	58.97 (58.63, 59.30)	3.01 (2.96, 3.05)

Table 1. *Cont.*

	All Ages	Age ≥ 55	Age < 55
		White	
Women			
Cases (%)	101,670 (49.3)	93,247 (45.2)	8400 (4.1)
Incidence Rate (95% CI)	11.11 (11.04, 11.18)	46.06 (45.76, 46.36)	3.08 (3.02, 3.14)
Men			
Cases (%)	104,731 (50.7)	92,927 (45.0)	11,792 (5.7)
Incidence Rate (95% CI)	14.6 (14.51, 14.69)	60.11 (59.72, 60.50)	2.21 (2.16, 2.26)
		Black	
Women			
Cases (%)	17,579 (53.6)	15,159 (46.2)	2412 (7.4)
Incidence Rate (95% CI)	14.41 (14.19, 14.62)	58.93 (57.99, 59.89)	3.17 (3.04, 3.30)
Men			
Cases (%)	15,216 (46.4)	12,482 (38.1)	2732 (8.3)
Incidence Rate (95% CI)	17.25 (16.96, 17.55)	69.83 (68.53, 71.14)	4.08 (3.93, 4.24)
		Hispanic	
Women			
Cases (%)	13,634 (51.9)	11,501 (43.8)	2114 (8.0)
Incidence Rate (95% CI)	10.65 (10.47, 10.84)	44.45 (43.63, 45.29)	2.01 (1.93, 2.10)
Men			
Cases (%)	12,647 (48.1)	10,162 (38.7)	2476 (9.4)
Incidence Rate (95% CI)	12.31 (12.08, 12.54)	51.13 (50.08, 52.19)	2.42 (2.32, 2.51)

Case percentages are reported as a percent of all cases; CI: confidence interval.

3.2. Older Adults (Age ≥ 55)

A total of 251,360 cases of PC were reported during the study period. Sex-specific incidence rates per 100,000 were 46.49 (95% CI: 46.23–46.75) in women and 58.97 (95% CI: 58.63–59.30) in men.

Incidence rates were relatively increasing at a lower rate in White women (AAPC = 0.80, 95% CI: 0.64–0.96, $p < 0.01$) compared to White men (AAPC = 1.12, 95% CI: 1.00–1.24, $p < 0.01$), with a statistically significant AAPC difference of −0.32 (95% CI: −0.52–−0.13, $p < 0.01$) (Table 2, Figure 1). These trends were neither identical ($p < 0.01$) nor parallel ($p < 0.01$). No sex differences in incidence rates were found in the Black and Hispanic groups.

Table 2. Pancreatic cancer incidence time trends for White, Black and Hispanic patients from 2000 to 2018 stratified by age and sex.

	White	*p*	Black	*p*	Hispanic	*p*
			All Ages			
Men AAPC (95% CI)	1.10 (0.99, 1.21)	<0.001	0.50 (0.25, 0.76)	<0.001	0.63 (0.18, 1.09)	0.009
Women ‡ AAPC (95% CI)	0.87 (0.74, 1.01)	<0.001	0.37 (0.10, 0.64)	0.011	0.68 (0.38, 0.99)	<0.001
ΔAAPC (95% CI)	−0.23 * (−0.39, −0.07)	0.005	−0.13 (−0.48, 0.22)	0.460	0.05 (−0.46, 0.56)	0.847

Table 2. *Cont.*

	White	p	Black	p	Hispanic	p
			Age ≥ 55			
Men AAPC (95% CI)	1.12 (1.00, 1.24)	<0.001	0.63 (0.36, 0.89)	<0.001	0.51 (0.01, 1.02)	0.046
Women ‡ AAPC (95% CI)	0.80 (0.64, 0.96)	<0.001	0.17 (−0.18, 0.51)	0.322	0.38 (0.05, 0.72)	0.027
ΔAAPC (95% CI)	−0.32 * (−0.52, −0.13)	0.001	−0.46 (−0.86, −0.06)	0.026	−0.13 (−0.69, 0.43)	0.648
			Age < 55			
Men AAPC (95% CI)	0.94 (0.61, 1.28)	<0.001	−0.22 (−1.01, 0.58)	0.569	1.48 (0.68, 2.82)	0.001
Women ‡ AAPC (95% CI)	1.68 (1.18, 2.17)	<0.001	1.74 (1.04, 2.46)	<0.001	3.03 (2.03, 4.04)	<0.001
ΔAAPC (95% CI)	0.73 * (0.18, 1.29)	0.010	1.96 * (0.98, 2.95)	<0.001	1.55 * (0.36, 2.75)	0.011

* AAPC trends used to calculate ΔAAPC are non-parallel and non-identical. ‡ Reference group when calculating ΔAAPC. Incidence trends were calculated per 100,000 people. AAPC: average annual percent change. CI: Confidence. Interval. ΔAAPC: difference in AAPC between women and men.

Incidence rates were also relatively higher in White women (AAPC = 0.80, 95% CI: 0.64–0.96, $p < 0.01$) compared to Hispanic women (AAPC = 0.38, 95% CI: 0.05–0.72, $p = 0.027$), with a difference in AAPC of 0.41 (95% CI: 0.06–0.76, $p = 0.021$) (Table 3, Figure 2). These trends were also neither identical ($p < 0.01$) or parallel ($p = 0.016$). Furthermore, incidence rates were relatively higher in White women (AAPC = 0.80, 95% CI: 0.64–0.96, $p < 0.01$) compared to Black women (AAPC = 0.17, 95% CI: −0.18–0.51, $p = 0.32$), with a difference in AAPC of 0.63 (95% CI: 0.27–0.99, $p < 0.01$). These trends were neither identical ($p < 0.01$) nor parallel ($p < 0.01$).

Table 3. Pancreatic cancer incidence time trends for women from 2000 to 2018 stratified by age, race and ethnicity as well as comparison of time trends between White, Black and Hispanic Women.

	All Ages	p	Age ≥ 55	p	Age < 55	p
White AAPC (95% CI)	0.87 (0.74, 1.01)	<0.001	0.80 (0.64, 0.96)	<0.001	1.68 (1.18, 2.17)	<0.001
Black AAPC (95% CI)	0.37 (0.10, 0.64)	0.011	0.17 (−0.18, 0.51)	0.322	1.74 (1.04, 2.46)	<0.001
Hispanic AAPC (95% CI)	0.68 (0.38, 0.99)	<0.001	0.38 (0.05, 0.72)	0.027	3.03 (2.03, 4.04)	<0.001
			Trend Comparison Among Race and Ethnicity Cohorts			
Black ‡ vs. Hispanic ΔAAPC (95% CI)	−0.31 (−0.69, 0.07)	0.108	−0.22 (−0.66, 0.23)	0.343	−1.28 (−2.42, −0.14)	0.028
White ‡ vs. Black ΔAAPC (95% CI)	0.50 * (0.22, 0.79)	<0.001	0.63 * (0.27, 0.99)	0.001	−0.07 (−0.87, 0.73)	0.866
White ‡ vs. Hispanic ΔAAPC (95% CI)	0.19 (−0.12, 0.50)	0.222	0.41 * (0.06, 0.76)	0.021	−1.35 (−2.39, −0.31)	0.011

* AAPC trends used to calculate differences in AAPC are non-parallel and non-identical. ‡ Reference group when calculating ΔAAPC. Incidence trends were calculated per 100,000 people. AAPC: average annual percent change. CI: confidence interval. ΔAAPC: difference in AAPC between women and men.

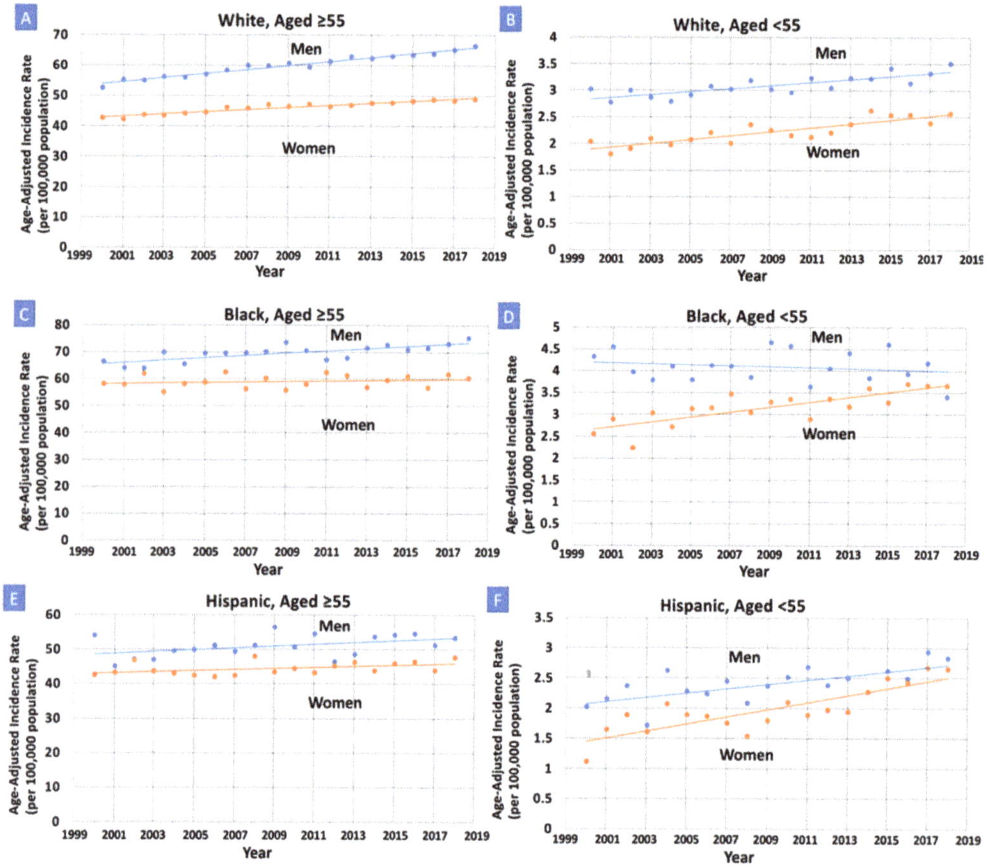

Figure 1. Sex-specific trends and age-adjusted pancreatic cancer incidence rates per 100,000 people among older (Age $\geq$ 55) and younger (Age < 55) White, Black and Hispanic adults. APC: annual percent change. (**A**): Older White: APC increasing at a lower rate in women compared to men (0.80 vs. 1.12, $p = 0.001$), with trends that are non-parallel ($p < 0.01$) and non-identical ($p < 0.01$). (**B**): Younger White: APC increasing at a greater rate in women compared to men (1.68 vs. 0.94, $p = 0.010$), with trends that are non-parallel ($p < 0.01$) and non-identical ($p < 0.01$). (**C**): Older Black: APC increasing at a lower rate in women compared to men (0.17 vs. 0.63, $p = 0.026$), with trends that are parallel ($p = 0.071$) and non-identical ($p < 0.01$). (**D**): Younger Black: APC increasing at a greater rate in women compared to men (1.74 vs. −0.22, $p < 0.001$), with trends that are non-parallel ($p < 0.01$) and non-identical ($p < 0.01$). (**E**): Older Hispanic: APC neither increasing or decreasing in women and men, no significant difference (0.38 vs. 0.51, $p = 0.65$). (**F**): Younger Hispanic: APC increasing at a greater rate in women compared to men (3.03 vs. 1.48, $p = 0.011$); trends are non-parallel ($p < 0.01$) and non-identical ($p < 0.01$).

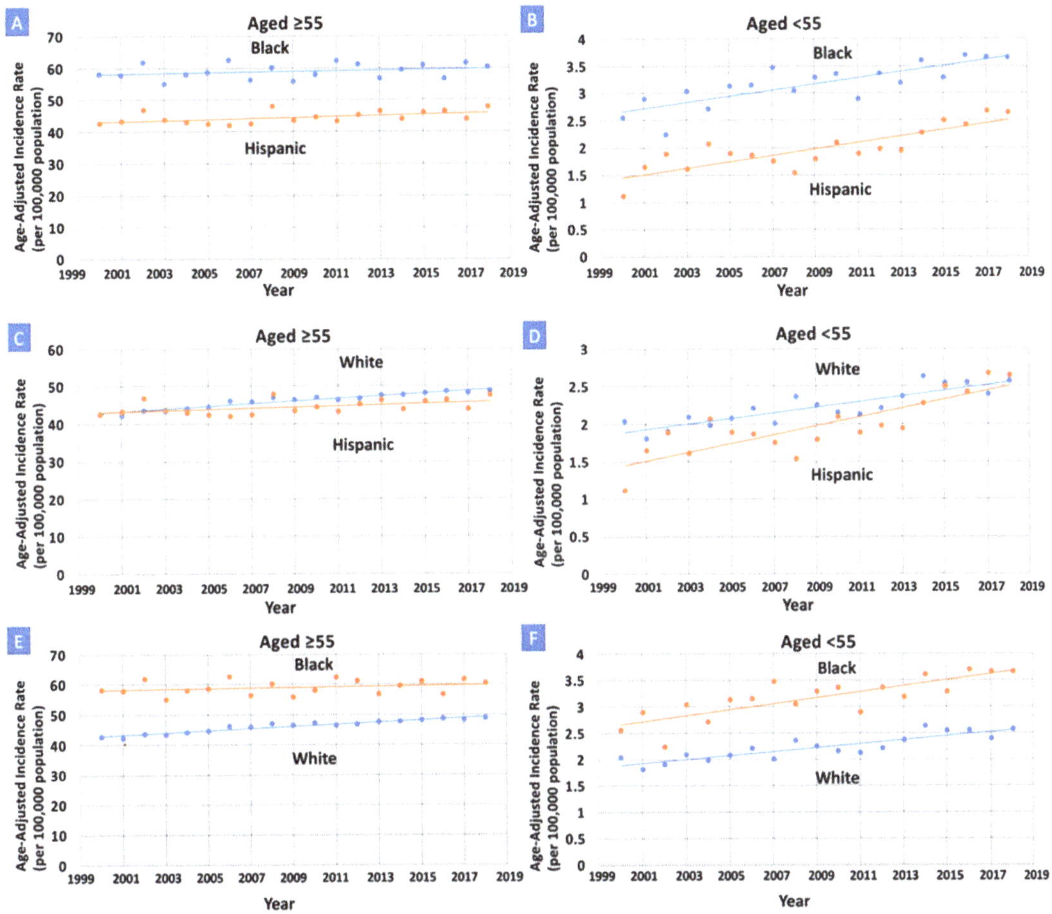

Figure 2. Race and ethnicity specific trends and age-adjusted pancreatic cancer incidence rates per 100,000 people among older (Age ≥ 55) and younger women (Age < 55). APC: annual percent change. (**A**): APC increasing in Hispanics and unchanged in Black older women; there is no significant difference (0.38 vs. 0.17, $p = 0.34$). (**B**): APC increasing at a greater rate in Hispanic compared to Black younger women (3.03 vs. 1.74, $p = 0.028$), with trends that are parallel ($p = 0.074$) and non-identical ($p < 0.01$). (**C**): APC increasing at a greater rate in White compared to Hispanic older women (0.80 vs. 0.38, $p = 0.021$), with trends that are non-parallel ($p = 0.016$) and non-identical ($p < 0.01$). (**D**): APC increasing at a greater rate in Hispanic compared to White younger women (3.03 vs. 1.68, $p < 0.001$), with trends that are parallel ($p = 0.070$) and non-identical ($p < 0.01$). (**E**): APC increasing at a greater rate in White compared to Black older women (0.80 vs. 0.17, $p = 0.001$), with trends that are non-parallel ($p < 0.01$) and non-identical ($p < 0.01$). (**F**): APC increasing in Black and White younger women, with no significant difference (1.74 vs. 1.68, $p = 0.87$).

4. Younger Adults (Ages < 55)

A total of 32,369 cases of PC were reported during the study period. Most cases occurred among men (6.5% of all cases), and the sex-specific incidence rates per 100,000 were 2.25 (95% CI: 2.21–2.28) in women and 3.01 (95% CI: 2.96–3.05) in men.

Incidence rates were relatively increasing at a greater rate in women compared to men in all three race and ethnic groups. White women (AAPC = 1.68, 95% CI: 1.18–2.17, $p < 0.01$) experienced a greater rate of increase in incidence rates compared to White men

(AAPC = 0.94, 95% CI: 0.61–1.28, $p < 0.01$) with a statistically significant AAPC difference of 0.73 (95% CI: 0.18–1.29, $p = 0.01$). These trends were neither identical ($p < 0.01$) nor parallel ($p < 0.01$). Black women (AAPC = 1.74, 95% CI: 1.04–2.46, $p < 0.01$) experienced an increase in incidence rate compared to Black men (AAPC = −0.22, 95% CI: −1.01–0.48, $p = 0.57$), with a statistically significant AAPC difference of 1.96 (95% CI: 0.98–2.95, $p < 0.01$). These trends were neither identical ($p < 0.01$) nor parallel ($p = 0.02$). Hispanic women (AAPC = 3.03, 95% CI: 2.03–4.04, $p < 0.01$) experienced a greater rate of increase in incidence rates compared to Hispanic men (AAPC = 1.48, 95% CI: 0.68–2.82, $p < 0.01$), with a statistically significant AAPC difference of 1.55 (95% CI: 0.36–2.75, $p = 0.011$). These trends were neither identical ($p < 0.01$) nor parallel ($p < 0.01$).

Incidence rates were relatively higher in Hispanic women (AAPC = 3.03, 95% CI: 2.03–4.04, $p < 0.01$) compared to Black women (AAPC = 1.74, 95% CI: 1.04–2.46, $p < 0.01$), with a difference in AAPC of −1.28 (95% CI: −2.42–−0.14, $p = 0.03$). These trends were non-identical ($p < 0.01$) and parallel ($p = 0.07$). Furthermore, incidence rates were relatively higher in Hispanic women (AAPC = 3.03, 95% CI: 2.03–4.04, $p < 0.01$) compared to White women (AAPC = 1.68, 95% CI: 1.18–2.17, $p < 0.01$), with a difference in AAPC of −1.35 (95% CI: − 2.39–−0.31, $p = 0.01$). These trends were non-identical ($p < 0.01$) and parallel ($p = 0.07$).

5. Discussion

We investigated our recent finding of an alarming increase in PC incidence rates in young women compared to young men by examining the association of race and ethnicity with these trends [4]. Initially, we examined sex disparity in PC incidence time trends stratified by age, race and ethnicity. While younger women (age < 55) showed a greater relative rate of increase in PC incidence compared to young men in all races and ethnic groups, rates disproportionately increased in younger Hispanic and Black women. Further analysis revealed that younger Hispanic women experienced significantly higher rates of increase in incidence compared to younger Black and White women. The present analysis builds on our previous findings by identifying racial and ethnic disparities in PC incidence, specifically in younger women. Given PC is expected to become the second leading cause of cancer death by the year 2030 in the US [17], we hope our findings aid in designing future studies and public health strategies to combat these alarming trends.

A racial gap in PC incidence has existed in the US since the 1970s with a greater incidence in Black compared to White patients, although known risk factors such as socioeconomic status (SES), lifestyle and biological variables do not alone independently explain these trends [18]. While our analysis did not investigate causative factors, the trends found in our analysis are likely due to changes in trends of modifiable as well as non-modifiable risk factors. SES has been shown to be associated with cancer incidence and outcomes, with the directionality of the relationship varying by cancer type [19]. A systematic review of European studies found variable effects of SES on PC incidence, varying by country of origin and, interestingly, patient sex [20]. Diabetes mellitus (DM), obesity and smoking have also been associated with PC incidence [21].

Obesity and DM have both been shown to disproportionately affect minority communities, although their impact on racial differences in PC incidence remains unclear [22,23]. An analysis of the NHANES surveys examined trends in the prevalence of diagnosed and undiagnosed DM in the United States from 1988 to 2012 and showed an increase in DM rates across all demographics, although interestingly only among those with a BMI of 30 or greater [24]. Furthermore, rates of undiagnosed DM only increased among Mexican American participants, with undiagnosed DM rates decreasing among all demographics except for Mexican Americans and the youngest age group, highlighting a possible disparity in access to care. A more recent study that examined trends in DM rates from 1999 to 2016 using the NHANES showed a significantly more dramatic increase in DM rates among Mexican Americans during the study period compared to the Non-Hispanic Black and Non-Hispanic White groups [25]. Interestingly, men experienced a greater increase in

DM rates overall compared to women. Obesity trends have also been explored through the NHANES, with recent studies finding an overall increase in obesity rates over the past few decades, with higher rates in younger cohorts compared to older [26,27]. While similar trend patterns of severe obesity were found on subgroup analysis by demographics, males were less likely to have severe obesity compared to females in all birth cohorts [27]. Lastly, a recent review shows smoking rates have decreased among all racial and ethnic groups in both men and women in recent decades, with Asian and Hispanic/Latino individuals demonstrating the lowest prevalence [28]. The review also highlights significant racial, ethnic and socioeconomic disparities in exposure to secondhand smoke, which may further drive disparities in risk. The literature is limited by a lack of simultaneous stratification of risk factors by race, ethnicity, sex and age. Furthermore, the relationship between the trends found in our analysis and these risk factors, along with other known and unknown risk factors, is complex and warrants further investigation.

The association of sex-specific factors with PC has also been investigated. A meta-analysis of 22 studies found higher parity was borderline 0.86 (95% CI: 0.73–1.02; $Q = 50.49$, $p < 0.001$, $I^2 = 58.4\%$) associated with a dose response decreased risk of PC when controlling for possible confounders such as smoking and diabetes mellitus rates [29]. It is proposed that the protective effect of pregnancy is likely sex hormone mediated. Estrogen has been shown in in vitro and in vivo studies to inhibit transplanted pancreatic carcinoma and the effects of early pancreatic carcinogenesis in rat models [30,31]. Furthermore, census data shows a decline in the overall birth rate in the US from 1980 to 2007 [32]. When stratified by race, birth rates significantly declined among Black women (84.9 to 72.7 births/1000 women) while birth rates mildly increased in White women (65.6 to 68.8 births/1000 women), although comparative analysis was not conducted between the races [32]. Unfortunately, birth rates for the Hispanic population were not available in this report.

The average age of diagnosis of PC is 71 years old, with up to 8% of patients diagnosed before the age of 50 [1,33]. Early-onset PC is relatively understudied, with an increasing incidence in recent years [33,34]. A recent analysis of 124,442 patients examined early and later age-onset PC incidence rates and found a higher percentage of men (58.3% vs. 49.8%), Black (16.8 vs. 12.2) and Hispanic (8.3 vs. 4.9) patients in the early group. Another study of 16,282 cases of PC also showed higher rates of Black (46%) and Hispanic (42%) patients diagnosed under the age of 65 compared to White (33%) and Asian (32%) patients ($p < 0.0001$) [35]. While early-onset PC has similar risk factors to late-onset PC, familial and genetic factors have been implicated as additional risk factors given that early-onset cases are more likely to have a family history of cancer and hereditary genetic syndromes [36]. Genetic factors are less likely to have had an impact on the change in trends revealed by our analysis due to the low likelihood of significant changes in the prevalence of genetic syndromes and hereditary diseases during our study period. Alternatively, we believe the disparity in incidence trends is likely due to changes in risk factor trends, both known and potentially unknown.

Limitations and Future Directions

The SEER database has certain limitations, including possible loss of records and coding reliability. Underreporting of cases as well as changes in testing rates may impact incidence rates across our study period. Furthermore, demographic data, including race and ethnicity, are often obtained from multiple sources, such as patient intake, provider notes, and administrative databases, allowing for misclassification bias and an impact on our findings. Given the self-reported nature of race and ethnicity, this may not correspond to genetic ancestry, which is an important consideration when drawing conclusions from our analysis. Migration of patients in and out of SEER registry areas and selection bias are also possible limitations that have been previously described [37]. Furthermore, due to the small sample size, the patient groups Asian/Pacific Islander, American Indian/Alaskan Native and Native Hawaiian were not included in our analysis. Our analysis is also limited by the lack of availabile data on risk factors that may explain these trends, making definitive conclusions regarding causal factors not possible. The relatively low 5-year survival rate of PC makes prevention through risk modification and effective screening guidelines crucial. Examining national trends in cancer incidence, such as in our study, highlights alarming trends warranting further investigation, ultimately leading to public health policy actions. We highly encourage future studies examining factors associated with the alarming disproportionate increase in incidence rates in young Black and Hispanic women compared to their male counterparts, as well as the disproportionate increase in younger Hispanic women compared to younger Black and White women. Given that race is a social construct, examination of trends stratified by race and ethnicity provides important insight into the impact of psychosocial, societal, and systemic factors on PC incidence and outcomes. To further improve screening and therapeutics, we encourage future investigation of genomic profiles as objective tools for improving screening, prevention, and therapeutics given the expected rise in incidence in the coming decade.

6. Conclusions

Younger women of all races and ethnicities experienced a greater rate of increase in pancreatic cancer incidence compared to their male counterparts; however, younger Hispanic and Black women experienced a disproportionately greater increase. Comparison between younger women revealed that younger Hispanic women experienced significantly higher rates of increase in incidence compared to younger Black and White women. Further studies are needed to better understand the factors associated with these disparities.

Supplementary Materials: The following supporting information can be downloaded at: https://www.mdpi.com/article/10.3390/cancers15030870/s1, Figure S1: Number of Pancreatic Cancer Cases During the Study Period (2000–2018) by Age Groups. Figure S2: Visual illustration of the test of Parallelism (Parallel Lines) and test of Coincidence (Identical Lines).

Author Contributions: J.S.S.: Substantial contributions to the conception or design of the work, interpretation of data for the work and drafting of the work, final approval of the version to be published; and agreement to be accountable for all aspects of the work in ensuring that questions related to the accuracy or integrity of any part of the work are appropriately investigated and resolved. Y.A.: acquisition, analysis, and interpretation of data for the work; and critical revision of the manuscript for important intellectual input. J.O.: Critical revision of the manuscript for important intellectual input. Y.J.: Critical revision of the manuscript for important intellectual input. S.J.P.: Critical revision of the manuscript for important intellectual input. R.W.: Critical revision of the manuscript for important intellectual input. K.P.: Critical revision of the manuscript for important intellectual input. Q.L.: Critical revision of the manuscript for important intellectual input. K.A.: Critical revision of the manuscript for important intellectual input. A.H.: Critical revision of the manuscript for important intellectual input. J.G.: Critical revision of the manuscript for important intellectual input. A.O.: Critical revision of the manuscript for important intellectual input. D.L.: Critical revision of the manuscript for important intellectual input. N.N.N.: Critical revision of the manuscript for important intellectual input. S.K.L.: Critical revision of the manuscript for important intellectual input. S.G.: Substantial contributions to the conception or design of the work, the acquisition, analysis and interpretation of data for the work; drafting of the work, critical revision of the work for important intellectual input; final approval of the version to be published; and agreement to be accountable for all aspects of the work in ensuring that questions related to the accuracy or integrity of any part of the work are appropriately investigated and resolved. All authors have read and agreed to the published version of the manuscript.

Funding: This project was supported in part by a philanthropic grant from The Widjaja Family Fund for Pancreatic Cancer Research. The funding agency did not have any role in the design and conduct of the study; the collection, management, analysis, and interpretation of the data; the preparation, review, or approval of the manuscript; and the decision to submit the manuscript for publication.

Institutional Review Board Statement: Given this data is publicly available and is de-identified, according to Cedars-Sinai institutional policy, this study was considered exempt from the institutional review board's (IRB) full or expedited review protocol. This is in accordance with the recommendations of the National Human Research Protections Advisory Committee.

Informed Consent Statement: Given that this data is publicly available and is de-identified, informed consent from study participants was not obtained.

Data Availability Statement: The data presented in this study are openly available at https://seer.cancer.gov/data-software/ (accessed on 22 January 2022).

Acknowledgments: The authors would like to thank Shelly Lu (Karsh Division of Gastroenterology and Hepatology, Cedars-Sinai Medical Center, Los Angeles, California, USA) for her mentorship, guidance, and support in all stages of this study.

Conflicts of Interest: The authors declare no conflict of interest.

References

1. Cancer Stat Facts: Pancreatic Cancer. Available online: https://seer.cancer.gov/statfacts/html/pancreas.html (accessed on 22 January 2022).
2. Soefje, S.A. Managing the economic impact of advanced pancreatic cancer. *Am. J. Manag. Care* **2019**, *25* (Suppl. S1), S11–S16. [PubMed]
3. Bauer, M.R.; Bright, E.E.; MacDonald, J.J.; Cleary, E.H.; Hines, O.J.; Stanton, A.L. Quality of Life in Patients with Pancreatic Cancer and Their Caregivers: A Systematic Review. Pancreas. *Pancreas* **2018**, *47*, 368–375. [CrossRef] [PubMed]
4. Gaddam, S.; Abboud, Y.; Oh, J.; Samaan, J.S.; Nissen, N.N.; Lu, S.C.; Lo, S.K. Incidence of Pancreatic Cancer by Age and Sex in the US, 2000-2018. *JAMA* **2021**, *326*, 2075. [CrossRef] [PubMed]
5. Ashktorab, H.; Kupfer, S.S.; Brim, H.; Carethers, J.M. Racial Disparity in Gastrointestinal Cancer Risk. *Gastroenterology* **2017**, *153*, 910–923. [CrossRef] [PubMed]
6. Tavakkoli, A.; Singal, A.G.; Waljee, A.K.; Elmunzer, B.J.; Pruitt, S.L.; McKey, T.; Rubenstein, J.H.; Scheiman, J.M.; Murphy, C.C. Racial Disparities and Trends in Pancreatic Cancer Incidence and Mortality in the United States. *Clin. Gastroenterol. Hepatol.* **2020**, *18*, 171–178.e10. [CrossRef]
7. Surveillance, Epidemiology, and End Results (SEER) Program. SEER*Stat Database: Incidence—SEER Research Limited-Field Data with Delay-Adjustment, 21 Registries, Malignant Only, Nov 2020 Sub (2000–2018)—Linked to County Attributes—Time Dependent (1990-2018) Income/Rurality, 1969–2019 Counties, National Cancer Institute, DCCPS, Surveillance Research Program, Released April 2021, Based on the November 2020 Submission. Available online: www.seer.cancer.gov (accessed on 22 January 2022).
8. Number of Persons by Race and Hispanic Ethnicity for SEER Participants (2020 Census Data). Available online: https://seer.cancer.gov/registries/data.html#a1 (accessed on 23 January 2022).
9. da Costa, W.L.; Oluyomi, A.O.; Thrift, A.P. Trends in the Incidence of Pancreatic Adenocarcinoma in All 50 United States Examined Through an Age-Period-Cohort Analysis. *JNCI Cancer Spectr.* **2020**, *4*, pkaa033. [CrossRef]
10. Siegel, R.L.; Miller, K.D.; Jemal, A. Colorectal Cancer Mortality Rates in Adults Aged 20 to 54 Years in the United States, 1970-2014. *JAMA* **2017**, *318*, 572. [CrossRef]
11. Joinpoint Regression Program, Version 4.9.0.0. March 2021. Available online: https://surveillance.cancer.gov/joinpoint/ (accessed on 22 January 2022).
12. Clegg, L.X.; Hankey, B.F.; Tiwari, R.; Feuer, E.J.; Edwards, B.K. Estimating average annual per cent change in trend analysis. *Stat. Med.* **2009**, *28*, 3670–3682. [CrossRef]
13. Number of Joinpoints. Joinpoint Help System. Available online: https://surveillance.cancer.gov/help/joinpoint/setting-parameters/method-and-parameters-tab/number-of-joinpoints (accessed on 9 November 2021).
14. Kim, H.-J.; Fay, M.P.; Yu, B.; Barrett, M.J.; Feuer, E.J. Comparability of Segmented Line Regression Models. *Biometrics* **2004**, *60*, 1005–1014. [CrossRef]
15. Kim, H.-J.; Fay, M.P.; Feuer, E.J.; Midthune, D.N. Permutation tests for joinpoint regression with applications to cancer rates. *Stat. Med.* **2000**, *19*, 335–351. [CrossRef]
16. *Joinpoint Regression Program VM*; Statistical Research and Applications Branch NCI: Bethesda, MD, USA, 2021.
17. Rahib, L.; Smith, B.D.; Aizenberg, R.; Rosenzweig, A.B.; Fleshman, J.M.; Matrisian, L.M. Projecting cancer incidence and deaths to 2030: The unexpected burden of thyroid, liver, and pancreas cancers in the United States. *Cancer Res.* **2014**, *74*, 2913–2921. [CrossRef] [PubMed]
18. Scarton, L.; Yoon, S.; Oh, S.; Agyare, E.; Trevino, J.; Han, B.; Lee, E.; Setiawan, V.W.; Permuth, J.B.; Schmittgen, T.D.; et al. Pancreatic Cancer Related Health Disparities: A Commentary. *Cancers* **2018**, *10*, 235. [CrossRef] [PubMed]
19. Singh, G.K.; Jemal, A. Socioeconomic and Racial/Ethnic Disparities in Cancer Mortality, Incidence, and Survival in the United States, 1950–2014: Over Six Decades of Changing Patterns and Widening Inequalities. *J. Environ. Public Health* **2017**, *2017*, 2819372. [CrossRef] [PubMed]

20. Mihor, A.; Tomsic, S.; Zagar, T.; Lokar, K.; Zadnik, V. Socioeconomic inequalities in cancer incidence in Europe: A comprehensive review of population-based epidemiological studies. *Radiol. Oncol.* **2020**, *54*, 1–13. [CrossRef] [PubMed]
21. Midha, S.; Chawla, S.; Garg, P.K. Modifiable and non-modifiable risk factors for pancreatic cancer: A review. *Cancer Lett.* **2016**, *381*, 269–277. [CrossRef] [PubMed]
22. Ogden, C.L.; Carroll, M.D.; Kit, B.K.; Flegal, K.M. Prevalence of Childhood and Adult Obesity in the United States, 2011–2012. *JAMA* **2014**, *311*, 806–814. [CrossRef] [PubMed]
23. Statistics about Diabetes. Available online: https://www.diabetes.org/resources/statistics/statistics-about-diabetes (accessed on 24 January 2022).
24. Menke, A.; Casagrande, S.S.; Geiss, L.S.; Cowie, C.C. Prevalence of and Trends in Diabetes among Adults in the United States, 1988–2012. *JAMA* **2015**, *314*, 1021. [CrossRef]
25. Fang, M. Trends in the Prevalence of Diabetes among U.S. Adults: 1999–2016. *Am. J. Prev. Med.* **2018**, *55*, 497–505. [CrossRef]
26. Li, M.; Gong, W.; Wang, S.; Li, Z. Trends in body mass index, overweight and obesity among adults in the USA, the NHANES from 2003 to 2018: A repeat cross-sectional survey. *BMJ Open* **2022**, *12*, e065425. [CrossRef] [PubMed]
27. Fan, K.; Lv, F.; Li, H.; Meng, F.; Wang, T.; Zhou, Y. Trends in obesity and severe obesity prevalence in the United States from 1999 to 2018. *Am. J. Hum. Biol.* **2022**, e23855. [CrossRef] [PubMed]
28. Drope, J.; Liber, A.C.; Cahn, Z.; Stoklosa, M.; Kennedy, R.; Douglas, C.E.; Henson, R.; Drope, J. Who's still smoking? Disparities in adult cigarette smoking prevalence in the United States: Disparities in Adult Smoking Prevalence in the US. *CA A Cancer J. Clin.* **2018**, *68*, 106–115. [CrossRef] [PubMed]
29. Guan, H.-B.; Wu, L.; Wu, Q.-J.; Zhu, J.; Gong, T. Parity and Pancreatic Cancer Risk: A Dose-Response Meta-Analysis of Epidemiologic Studies. *PLoS ONE* **2014**, *9*, e92738. [CrossRef] [PubMed]
30. Sumi, C.; Longnecker, D.S.; Roebuck, B.D.; Brinck-Johnsen, T. Inhibitory effects of estrogen and castration on the early stage of pancreatic carcinogenesis in Fischer rats treated with azaserine. *Cancer Res.* **1989**, *49*, 2332–2336.
31. Sumi, C.; Brinck-Johnsen, T.; Longnecker, D.S. Inhibition of a transplantable pancreatic carcinoma by castration and estradiol administration in rats. *Cancer Res.* **1989**, *49*, 6687–6692.
32. Table 80. Births, Birth Rates, and Fertility Rates by Race, Sex, and Age: 1980 to 2007. Available online: https://www2.census.gov/library/publications/2010/compendia/statab/130ed/tables/11s0080.pdf (accessed on 24 January 2022).
33. Alese, O.B.; Jiang, R.; Shaib, W.; Wu, C.; Akce, M.; Gaines, T.; Ni, L.; Behera, M.; El-Rayes, B.F. Young Adults with Pancreatic Cancer: National Trends in Treatment and Outcomes. *Pancreas* **2020**, *49*, 341–354. [CrossRef] [PubMed]
34. Sung, H.; Siegel, R.L.; Rosenberg, P.S.; Jemal, A. Emerging cancer trends among young adults in the USA: Analysis of a population-based cancer registry. *Lancet Public Health* **2019**, *4*, e137–e147. [CrossRef] [PubMed]
35. Singal, V.; Singal, A.K.; Kuo, Y.-F. Racial disparities in treatment for pancreatic cancer and impact on survival: A population-based analysis. *J. Cancer Res. Clin. Oncol.* **2012**, *138*, 715–722. [CrossRef]
36. Primavesi, F.; Stättner, S.; Schlick, K.; Kiesslich, T.; Mayr, C.; Klieser, E.; Urbas, R.; Neureiter, D. Pancreatic cancer in young adults: Changes, challenges, and solutions. *OncoTargets Ther.* **2019**, *12*, 3387–3400. [CrossRef]
37. Yu, J.B.; Gross, C.P.; Wilson, L.D.; Smith, B.D. NCI SEER public-use data: Applications and limitations in oncology research. *Oncology* **2009**, *23*, 288–295.

cancers

MDPI

Article

Gastrointestinal Adenocarcinoma Incidence and Survival Trends in South Australia, 1990–2017

Dominique Schell [1], Shahid Ullah [1], Mark E. Brooke-Smith [2], Paul Hollington [1,2], Marina Yeow [2], Christos S. Karapetis [1,3,4], David I. Watson [1,2,3], Stephen J. Pandol [5], Claire T. Roberts [1,3,*] and Savio G. Barreto [1,2,3,*]

1 College of Medicine and Public Health, Flinders University, Adelaide, SA 5042, Australia; sche0142@flinders.edu.au (D.S.); shahid.ullah@flinders.edu.au (S.U.); paul.hollington@sa.gov.au (P.H.); chris.karapetis@sa.gov.au (C.S.K.); david.watson@flinders.edu.au (D.I.W.)
2 Flinders Medical Center, Division of Surgery and Perioperative Medicine, Flinders University, Adelaide, SA 5042, Australia; mark.brooke-smith@flinders.edu.au (M.E.B.-S.); marina.yeow2@sa.gov.au (M.Y.)
3 Flinders Health and Medical Research Institute, Flinders University, Adelaide, SA 5042, Australia
4 Department of Medical Oncology, Flinders Medical Centre, Flinders University, Adelaide, SA 5042, Australia
5 Cedars-Sinai Medical Center, Division of Digestive and Liver Diseases, Los Angeles, CA 90048, USA; stephen.pandol@cshs.org
* Correspondence: Claire.roberts@flinders.edu.au (C.T.R.); savio.barreto@sa.gov.au (S.G.B.)

Simple Summary: This study from South Australia using the state's Cancer Registry data provides compelling evidence for a significant increase in the incidence of young-onset (18–50 years) gastrointestinal (oesophageal, stomach, colon and rectum, and pancreas) adenocarcinomas over the last three decades. The trend observed in the young cohort was not mirrored in older individuals >50 years. This increased incidence, though noted in both sexes, was more pronounced in males compared to females. Survival in the young-onset adenocarcinoma cohort was only seen in patients with colorectal cancers, but not oesophagus, stomach and pancreas. This study calls for a concerted effort aimed at determining the sociodemographic factors underlying this disturbing trend with the aim of developing preventative strategies.

Abstract: Background & Aims: Globally, there has been a concerning rise in the incidence of young-onset cancers. The aim of this study was to provide trends in the incidence and survival of gastrointestinal adenocarcinomas (oesophagus, stomach, pancreas, and colorectal) in South Australia over a 27-year period. Methods: This is a cross-sectional analysis of a prospective longitudinal database including all cases of gastrointestinal adenocarcinomas prospectively reported to the South Australian (State) Cancer Registry from 1990 to 2017. Results: A total of 28,566 patients diagnosed with oesophageal, stomach, pancreatic, or colorectal adenocarcinoma between 1990 and 2017 were included in the study. While the overall incidence for gastrointestinal adenocarcinomas in individuals >50 years has decreased since 2000 (IRR of 0.97 (95% CI 0.94–1.00; $p = 0.06$)) compared to 1990–1999, the rate amongst individuals aged 18–50 has significantly increased (IRR 1.41 (95% CI 1.27–1.57; $p < 0.001$)) during the same reference time period. Although noted in both sexes, the rate of increase in incidence was significantly greater in males (11.5 to 19.7/100,000; $p < 0.001$). The overall survival from adenocarcinomas across all subsites improved in the >50-year cohort in the last decade (HR 0.89 (95% CI 0.86–0.93; $p < 0.001$)) compared to 1990–1999. In individuals aged 18–50 years, there has only been a significant improvement in survival for colorectal cancer (HR 0.82 (95% CI 0.68–0.99; $p < 0.04$)), but not the other subsites. A lower overall survival was noted for males in both age cohorts (18–50 years—HR 1.24 (95% CI 1.09–1.13; $p < 0.01$) and >50 years—HR 1.13 (95% CI 1.10–1.16; $p < 0.001$), respectively) compared to females. Conclusions: This study from South Australia demonstrates a significant increase in young-onset gastrointestinal adenocarcinomas over the last 28 years, with a greater increase in the male sex. The only significant improvement in survival in this cohort has been noted in colorectal cancer patients.

Citation: Schell, D.; Ullah, S.; Brooke-Smith, M.E.; Hollington, P.; Yeow, M.; Karapetis, C.S.; Watson, D.I.; Pandol, S.J.; Roberts, C.T.; Barreto, S.G. Gastrointestinal Adenocarcinoma Incidence and Survival Trends in South Australia, 1990–2017. *Cancers* **2022**, *14*, 275. https://doi.org/10.3390/cancers14020275

Academic Editors: Irit Ben Aharon and David Wong

Received: 2 December 2021
Accepted: 5 January 2022
Published: 6 January 2022

Keywords: outcomes; morbidity; mortality; stomach; pancreas; colon

1. Introduction

Globally, there have been several reports of increasing incidence of early- or young-onset cancers [1]. These encompass a spectrum of solid organ cancers such as colorectal cancer (CRC) and adenocarcinomas of the pancreas [2], breast [3], ovary [4], oesophagus [5], and stomach [6]. Though CRC has demonstrated the most striking trend, similar patterns have been uncovered in other gastrointestinal adenocarcinomas. In an effort to explain this disturbing trend, clinicians and scientists have attempted to implicate early life exposures to antibiotics [7], the impact of the gut microbiome [8], and variations in mismatch repair (MMR) genes and microsatellite instability (MSI) [9]. Siegel et al. [10] linked the trend to birth cohorts and the possible influence of the obesity pandemic (drawing on the evidence relating obesity, unhealthy eating habits, and sedentary lifestyles [11]), whilst Lui et al. [12] postulated the role of lifestyle factors such as physical inactivity and alcohol consumption in the development of early onset cancers. We proposed a hypothesis, with supporting evidence [13], drawing attention to the significant contribution of perinatal events [1] drawing on the work of Barker [14], Knudson [15], and Lahouel [16]. It was postulated [1,13] that an *'in utero'* insult to the foetus constitutes the 'first hit'. The second hit would then be the result of exposures occurring in childhood and adolescence.

Koczwara et al. [17] have recently demonstrated that the coexistence of comorbidities (diabetes mellitus, chronic pulmonary disease, cardio- and cerebrovascular diseases, and peptic ulcer disease) alongside a cancer diagnosis significantly worsens survival in younger individuals. This concerning observation strengthens the relationship between the 'developmental origins of health and disease (DOHaD)' [14] and our hypotheses [1,13,18]. It is, thus, imperative that the cause for young-onset carcinogenesis is further investigated with the aim of detecting, and hopefully correcting, the underlying factor(s) adversely affecting survival in this cohort. In South Australia, trends in the incidence, and/or survival, of young-onset adenocarcinomas affecting abdominal viscera have not been determined. South Australia is one of the six states of Australia. It is inhabited by a population of 1.77 million that is predominantly Caucasian (>80%) and reflective of diversity seen in European countries.

The aim of this study was to determine trends in the incidence and survival of gastrointestinal adenocarcinomas (oesophagus, stomach, pancreas, and colorectal) in South Australia over a 27-year period. Being empowered with this information will offer us the opportunity to determine if data generated on young-onset cancer from global research can be extrapolated to our population. It will also enable us to join the global efforts in deciphering the causes that underpin the development of young-onset adenocarcinomas, offering South Australians and Australians, at large, the hope of improving outcomes in cancer.

2. Methods

We performed a cross-sectional analysis of a prospective longitudinal database to include all cases of adenocarcinoma of the oesophagus, stomach, pancreas, colon, and rectum reported to the South Australian Cancer Registry since 1 January 1990, to the latest available date at the time of our analysis, 31 December 2017. All cases of invasive cancer are notifiable under the Cancer Reporting Regulations under the *South Australian Health Care Act 2008* [19]. The South Australian Cancer Registry was established in 1977. It is managed by Wellbeing SA's Epidemiology Branch (under the auspices of SA Health). The registry has several processes that enhance the quality of the data collection, such as electronic notification, a series of internal data checks prior to reporting the data, notification from multiple sources, and an annual internal deduplication procedure. Several features inbuilt in the Registry Plus software system also allow real time query. Statistics on quality measures can be found in the annual South Australian Cancer Registry reports [20]. The South Australian Cancer Registry has been able to completely capture all cancers across the time frame contributing to the strength of the data source.

Ethics approval for the study was obtained from the South Australian Department for Health and Wellbeing Human Research Ethics Committee (HREC) (reference number: LNR/21/SAC/51). Since only de-identified data was provided to us by the South Australian Cancer Registry, a waiver of consent was provided by the Ethics Committee.

2.1. Selection of Cases

2.1.1. Inclusion Criteria

All individuals who were aged 18 years and over and had a pathologically confirmed diagnosis of adenocarcinoma (ICD 10 codes: oesophagus = C15, stomach = C16, pancreas = C25, colon = C18, rectosigmoid junction = C19, rectum = C20). Histology codes for adenocarcinoma: ICD 8140/2, 8140/3, 8141/3, 8143/3, 8210/2, 8210/3, and 8230/2 and diagnosed in SA from 1 January 1990 to 31 December 2017 were included in this study. The study period was categorised into 3 eras (1990–1999, 2000–2009, 2010–2017) to reflect incidence and survivals of cancers over time. Some of the data for pancreatic adenocarcinoma has been previously published by us [21].

2.1.2. Statistical Analysis

All statistical analyses were conducted using R version 4.1.0 and Stata version 16.1. Patients' characteristics were expressed as median and interquartile range (IQR) for skewed data. The Mann–Whitney U test was used to explore the significance of differences in patients' age between two groups of patients. Proportions were presented as percentages of the respective denominator and were compared between groups using a standard chi-square test for association with continuity correction, where appropriate.

The incidence rates were calculated by taking the total number of cases divided by the population at risk. The rates were presented per 100,000 persons over 3 time periods for age groups 18–50 years and >50 years for each sex and cancer primary sites. A Poisson regression model was applied to examine the incidence rates between the groups of the above characteristics. The estimates were calculated using the likelihood ratio method and were expressed as incidence rate ratios (IRRs) from the Poisson model. Poisson regression model was also used to calculate the average annual percentage change.

Survival was measured from the date of cancer diagnosis to the date of death, and individuals were censored at date of loss to follow-up or census date. The census date was assigned on 31 December 2017. The South Australian Cancer Registry data are linked to the births, deaths, and marriage data once a year, in general. Cox proportional hazard models were applied to examine the survival outcomes. Sex, primary sites, and cohort era were used to explore the risk of death between two cohorts (18–50 years and >50 years). The estimates were calculated using the likelihood ratio method and were expressed as hazard ratios (HRs)—the lower the HR, the longer the survival. Proportional hazard assumption was tested by the log–log plot of survival and Schoenfeld residuals. Survival curves for patient survival were evaluated by standard Kaplan–Meier survival curves and patient cohorts were compared by log-rank test. The two-sided test was performed for all analysis, 95% confidence intervals were reported, and the level of significance was set at $p < 0.05$.

3. Results

3.1. Demographic Data and Time Trend of Reported Cases

A total of 28,566 patients were diagnosed with oesophageal, stomach, pancreatic, or colorectal adenocarcinoma in South Australia between 1990 and 2017 (2129, 7.5% patients aged 18–50 years and 26,437, 92.5% patients aged >50 years). The median ages for the 18–50 years and >50 years cohorts were 46 years (IQR 41–49 years) and 72 years (IQR 64–79 years), respectively (Table 1). Adenocarcinomas of the colon and rectum were the most common cancers in both age cohorts (8.85/100,000 for individuals aged

18–50 years and 159.27/100,000 for those aged >50 years). Age, in itself, was a contributing factor for incidence, with a higher increment for individuals aged 18–50 years (IRR = 1.17 (95% CI 1.16–1.18, p <0.001)) as compared to those >50 years (IRR 1.05 (95%CI 1.05–1.05, p <0.001)) (Table 2). The overall cancer incidence rates varied by sex within the two cohorts (9.42/100,000 for females and 11.78/100,000 for males aged 18–50 years and 156.19/100,000 for females and 242.33/100,000 for males >50 years). Both sexes in age cohort 18–50 years have experienced a significant increase in the incidence of adenocarcinomas (Figure 1; Supplementary Figure S1—depicting trends over 4-year intervals) over the three eras (females 8.3 to 11.9/100,000; p <0.001 and males 11.5 to 19.7/100,000; p <0.001). However, the incidence rates were significantly greater for males compared to females in both age cohorts, *viz.* 18–50 years (IRR 1.25 (95%CI 1.15–1.36; p <0.001)) and >50 years (IRR 1.55 (95%CI 1.51–1.59; p <0.001)) (Table 2). The incidence rates increased by 1% for every increment of year for males in age groups 18–50 years. However, no similar trend was noted for females in the same age group (Supplementary Table S1). The sex-specific incidence rates for gastrointestinal adenocarcinomas in individuals >50 years have reduced over the three eras (Figure 1). This significant trend persisted for every cancer site across both age cohorts (Figure 2, Supplementary Table S2).

Table 1. Patient's characteristics, era, and primary sites of cancer between two age groups (N = 28,566).

Patient Characteristics	(18–50 Years) n = 2129 (7.5%)		(>50 Years) n = 26,437 (92.5%)		
	n	%	n	%	p-Value
Age (years): median ± IQR	46	(41–49)	72	(64–79)	<0.001
Sex					0.09
Female	1190	55.9	15275	57.8	
Male	939	44.1	11162	42.2	
Era					0.49
1990–1999	650	30.5	7994	30.2	
2000–2009	720	33.8	9270	35.1	
2010–2017	759	35.7	9173	34.7	
Primary site					0.058
Colon and Rectum	1776	83.4	21422	81.0	
Pancreas	150	7.0	2163	8.2	
Stomach	127	6.0	1808	6.8	
Oesophagus	76	3.6	1044	3.9	

Note. Number and percentages are reported unless stated otherwise; IQR—interquartile range. The p values are based on Mann-Whitney U test for medians and chi-square test for proportions.

Table 2. Incidence rates (IR) and incidence rate ratios (IRR) for age, sex, era, and cancer sites between two age groups (n = 28,566).

Patient Characteristics	(18–50 Years) n = 2129			(>50 Years) n = 26,437		
	* IR (95% CI)	IRR (95% CI)	p-Value	* IR (95% CI)	IRR (95% CI)	p-Value
Overall	10.60 (10.16–11.06)			196.56 (194.20–198.94)		
Age (years)	-	1.17 (1.16–1.18)	<0.001	-	1.05 (1.05–1.05)	<0.001
Sex						
Female	9.42 (8.82–10.04)	Reference	-	156.19 (153.30–159.11)	Reference	-
Male	11.78 (11.12–12.46)	1.25 (1.15–1.36)	<0.001	242.33 (238.51–246.21)	1.55 (1.51–1.59)	<0.001
Era						
1990–1999	9.13 (8.44–9.86)	Reference	-	203.04 (198.61–207.53)	Reference	-
2000–2009	10.19 (9.46–10.96)	1.12 (1.00–1.24)	0.04	190.74 (186.87–194.66)	0.94 (0.91–0.97)	<0.001
2010–2017	12.89 (11.98–13.83)	1.41 (1.27–1.57)	<0.001	197.16 (193.15–201.24)	0.97 (0.94–1.00)	0.06
Cancer site						
Colon & Rectum	8.85 (8.44–9.27)	-	-	159.27 (157.15–161.42)	-	-
Pancreas	0.75 (0.63–0.88)	-	-	16.08 (15.41–16.77)	-	-
Stomach	0.63 (0.53–0.75)	-	-	13.44 (12.83–14.08)	-	-
Oesophagus	0.38 (0.30–0.47)	-	-	7.76 (7.30–8.25)	-	-

* IR is incidence per 100,000 South Australian residents. IRs were not reported for age and IRRs were not reported for cancer sites.

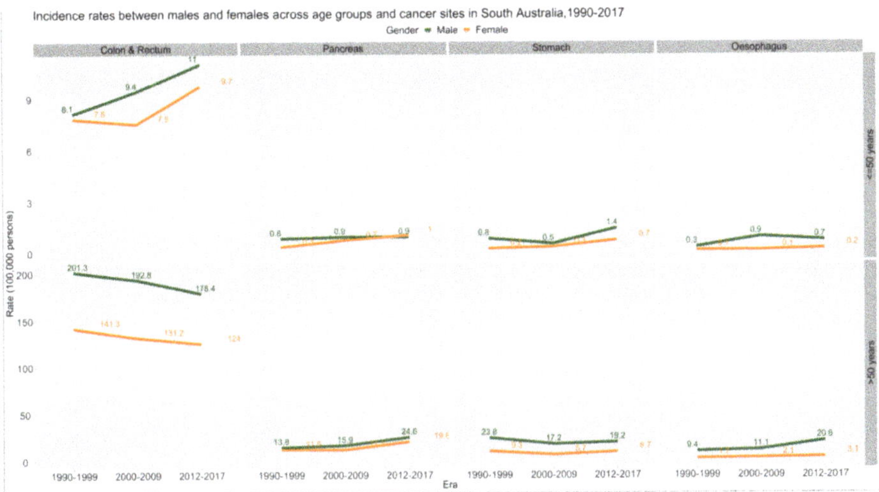

Figure 1. Trend in incidence rates by sex and era between two age groups across cancer sites 1990–2017 (n = 28,566).

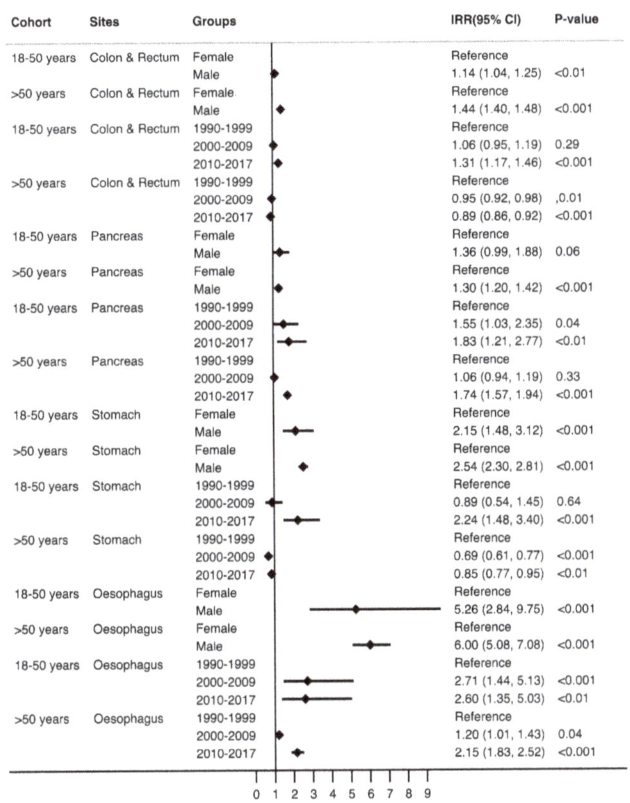

Figure 2. Incidence rate ratios (IRR) and 95% CI (Poisson regression model) for sex and era by primary sites between two age groups (n = 28,566).

3.2. Trends in the Incidence of Gastrointestinal Adenocarcinomas

While the overall incidence rate for gastrointestinal adenocarcinomas in individuals >50 years in South Australia has reduced over the last three decades from 1990–1999 to 2010–2017 (203.04/100,000 to 197.16/100,000), the rates amongst individuals aged 18–50 significantly increased over the same time period from (9.13/100,000 to 12.89/100,000) (Table 2). This was confirmed by the significantly increasing trend in the IRR of 1.41 (95% CI 1.27–1.57; $p < 0.001$) in individuals aged 18–50 years, compared to a declining trend in the IRR of 0.97 (95% CI 0.94–1.00; $p = 0.06$) in those >50 years. While the IRRs for colorectal and stomach cancer significantly decreased (0.89, 95% CI 0.86–0.92; $p < 0.001$ and 0.85, 95% CI 0.77–0.95; $p < 0.01$) in the most recent decade, for the cohort aged >50 years, the IRRs for pancreatic and oesophageal cancer demonstrated a significantly increased trend (1.74, 95% CI 1.57–1.94; $p < 0.001$ and 2.15, 95% CI 1.83–1.52; $p < 0.001$, respectively). For the young-onset cohort, however, every cancer site demonstrated a significantly increased trend in the IRR (oesophagus, 2.60, 95% CI 1.35–5.03; $p < 0.01$; stomach, 2.24, 95% CI 1.48–3.40; $p < 0.001$; pancreas, 1.83, 95% CI 1.21–2.77; $p < 0.01$, and colon and rectum, 1.31, 95% CI 1.17–1.46; $p < 0.001$) (Figure 2, Supplementary Table S2).

3.3. Survival by Time Trends and Site

The Kaplan–Meier survival estimates showed the greatest median survivals for colorectal cancer in both age cohorts (25.86 for individuals aged 18–50 years and 7.00 years for those >50 years (Table 3), and these were significantly better than the reference (oesophageal adenocarcinoma) (HR 0.28 (95% CI 0.22–0.37; $p < 0.001$) for those aged 18–50 years, and 0.34 (95% CI 0.31–0.36; $p < 0.001$) for those >50 years, respectively) (Table 4, Supplementary Figure S2). The longest median survival for colorectal cancer in the 18–50 years cohort was due to those patients diagnosed with colorectal cancer in the first half of the 1990–1999 era. The overall survival from adenocarcinomas across all subsites has improved significantly for the age cohort >50 years. However, despite demonstrating a trend in improvement, the result was not statistically significant for individuals in the 18–50 years cohort in the last decade (HR 0.92 (95% CI 0.79–1.08; $p = 0.32$)). The overall survival in males is significantly lower compared to females in both age cohorts (HR 1.24 (95% CI 1.09–1.40; $p < 0.01$) and 1.13 (95% CI 1.10–1.16; $p < 0.001$), respectively). This latter observation was largely the effect of stomach cancer (HR 1.74 (95% CI 1.09–2.80; $p = 0.02$)) in individuals aged 18–50 years and colorectal cancer (HR 1.11 (95% CI 1.07–1.15; $p < 0.001$)) in those aged >50 years (Figure 3, Supplementary Table S3). Males aged >50 years had a significantly better survival for oesophageal adenocarcinoma compared to females (HR 0.82 (95% CI 0.68–0.98; $p = 0.03$)). Reassuringly, the survival of individuals >50 years affected by most adenocarcinomas has significantly improved in the most recent decade (oesophagus, 0.83, 95% CI 0.70–0.98; $p = 0.03$; pancreas, 0.69, 95% CI 0.62–0.77; $p < 0.001$, and colon and rectum, 0.75, 95% CI 0.71–0.78; $p < 0.001$). However, in individuals aged 18–50 years, there has only been a significant improvement in survival following colorectal cancer (HR 0.82 (95% CI 0.68–0.99; $p = 0.04$)) (Figure 3, Supplementary Table S3).

Table 3. Median survival times for primary sites of cancer between two age groups ($n = 27,855$).

Cancer Sites	(18–50 Years)	(>50 Years)
	$n = 2107$ (7.6%)	$n = 25,748$ (92.4%)
	Median (95% CI)	Median (95% CI)
Overall	12.67 (9.19–17.07)	4.60 (4.44–4.77)
Colon and Rectum	25.86 (19.90-NA)	7.00 (6.81–7.24)
Pancreas	0.70 (0.56–0.86)	0.48 (0.44–0.52)
Stomach	1.32 (1.02–1.91)	0.94 (0.85–1.02)
Oesophagus	1.36 (0.87–2.54)	0.99 (0.88–1.05)

Note: Upper confidence level of survival time for colon and rectum cancer exceeded the follow-up time. NA = not available

Table 4. Hazard ratios (HR) and 95% CI (Cox proportional hazard model) for age, sex, era, and cancer sites between two age groups (n = 27,855).

Variables	(18–50 Years) n = 2107		(>50 Years) n = 25,748	
	HR (95% CI)	***p* Value**	**HR (95% CI)**	***p* Value**
Age (years)	1.01 (1.00–1.02)	0.03	1.04 (1.04–1.04)	<0.001
Sex				
Female	Reference	-	Reference	-
Male	1.24 (1.09–1.40)	<0.01	1.13 (1.10–1.16)	<0.001
Era				
1990–1999	Reference	-	Reference	-
2000–2009	0.86 (0.74–1.00)	0.046	0.85 (0.82–0.88)	<0.001
2010–2017	0.92 (0.79–1.08)	0.32	0.89 (0.86–0.93)	<0.001
Cancer site				
Colon and Rectum	0.28 (0.22–0.37)	<0.001	0.34 (0.31–0.36)	<0.001
Pancreas	2.48 (1.83–3.36)	<0.001	2.13 (1.97–2.31)	<0.001
Stomach	0.95 (0.69–1.32)	0.77	0.94 (0.86–0.36)	0.12
Oesophagus	Reference	-	Reference	-

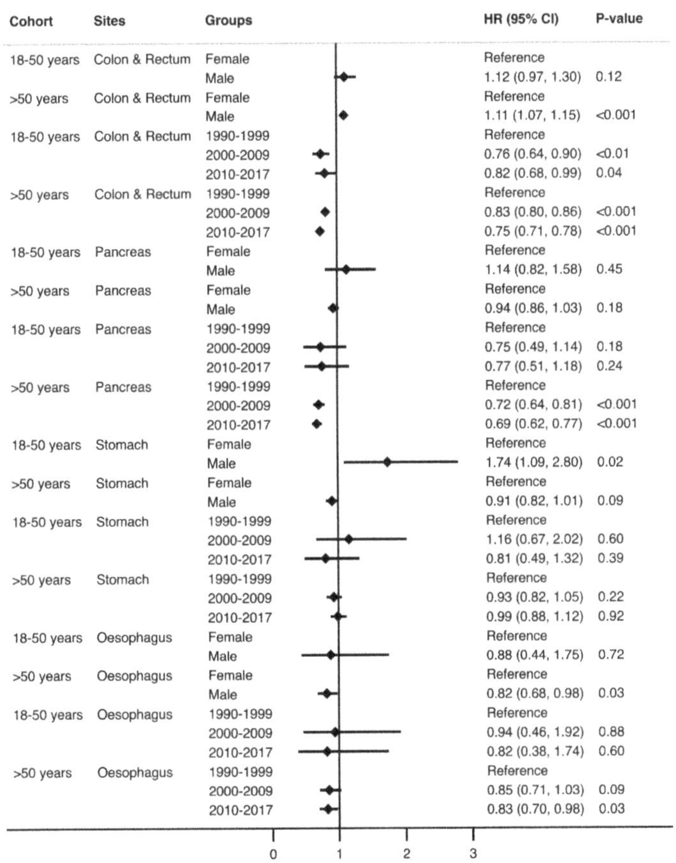

Figure 3. Hazard ratios (HR) and 95% CI (Cox proportional hazard model) for sex and era by primary sites between two age groups (n = 27,855).

4. Discussion

This study from South Australia demonstrates a rising incidence of young-onset (18–50 years) oesophagus, stomach, pancreas, and colorectal adenocarcinomas over a 28-year period despite a declining overall trend for individuals >50 years. In both age cohorts, the incidence rate ratio is significantly greater in males compared to females. The overall survival from adenocarcinomas across all subsites has improved significantly for the age cohort >50 years. In individuals aged 18–50 years, there has only been a significant improvement in survival for colorectal cancer, but not the other subsites.

This study presents a disturbing trend in the incidence of young-onset gastrointestinal adenocarcinomas in South Australia mirroring international data [10,12,22,23]. The change in incidence rates in younger individuals appears to be greatest for colorectal adenocarcinoma. A similar finding was previously reported by Feletto et al. [24] when studying cancers of the colon and rectum in Australia from 1982 to 2014 and by Young et al. [25] in 2015. Though the underlying causes remain to be deciphered, a similar trend in international studies has been postulated to be due to increases in early-life antibiotic use, obesity, the consumption of processed foods and alcohol [10,26,27], as well as an increase in metabolic disease (especially obesity and type 2 diabetes mellitus) seen in this similar time frame [28,29]. The relationship between the use of antibiotics and the risk of young-onset colorectal cancer has been studied by Zhang et al. [30] They found that antibiotic use was associated with a dose-dependent risk of colorectal cancer, and the location of the cancer (most commonly, proximal colon with the use of antibiotics with an anti-anaerobic activity, although an inverse relationship was noted in the rectum) altered depending on the type of antibiotic (penicillins increased the risk of colon cancer while tetracyclines were associated with a decreased risk of rectal cancer). The significant decline in colorectal adenocarcinomas in individuals >50 years in the same period of study is most likely due to early detection and prompt management of colorectal adenomas due to more effective screening of people in this age group since 2006 [24,31].

The evidence in the literature regarding the role of sex in young-onset gastrointestinal adenocarcinoma has been mixed, with some studies reporting a greater incidence rate in males for oesophageal, stomach, and colon and rectal cancers [32–34], while others reporting females showing greater incidence rates for colon cancer [35]. This study has demonstrated that the incidence of young-onset gastrointestinal adenocarcinomas has significantly increased in males and females between the ages of 18 and 50 years, with the increase being more pronounced in males. Putative explanations for individual adenocarcinomas include early-onset pancreatic cancer and higher rates of smoking in males [36], and early-onset stomach cancer and increased occurrence of *Helicobacter pylori* in males [33]. In the case of South Australia, given that rates for smoking have significantly reduced [37], while the prevalence of *Helicobacter pylori* infection has either remained stable or is on the decline in Australia [38,39], these are not tenable as contributory factors to the trend being witnessed. Nevertheless, although several studies report significant findings relating to sex, the reasoning for why the incidence is higher in males could possibly reflect the contribution of male central adiposity as compared to the subcutaneous adiposity that predominates in women.

While colorectal cancer had the greatest increase in incidence in the 18–50 years cohort globally, as well as in our study, the cause for this is still unclear. Family history has always been deemed a major risk factor in the development of colorectal cancer, but its role in young-onset carcinogenesis is less well understood. Although O'Connell et al. found that 22.7% of young-onset colorectal cancer had a positive family history for the disease [40], studies by Lee et al. and Dozois et al. revealed that early-onset colorectal cancer was mostly diagnosed in patients with no familial history and no genetic risk factors [41,42]. Similarly, it has been reported that the distribution of the tumour site differs significantly between those with a family history of colorectal cancer and those without, with the proximal colon being associated with patients with a positive family history, and distal colorectum for patients with no family history [43], suggesting a different carcinogenic mechanism altogether.

Mirroring these reports, Bergquist et al. [44] stated that the hereditary component of stomach cancer only accounts for a minority of young-onset stomach cancer, and Piciucchi et al. and Ntala et al. reported the same with regards to young-onset pancreatic cancer [45,46]. Hypothesized mechanisms lie in genetic susceptibilities expressed in single-nucleotide instability, somatic gene mutations and epigenetic alterations, [44] as well as environmental factors such as increased sedentary living and declining dietary quality in the past three decades [27]. Oesophageal adenocarcinoma has also been on the rise since 1990, with hypotheses focusing on the role of obesity and its significant link to the pathogenesis of Barrett's oesophagus [47], causing proinflammatory cytokines produced by visceral fat to promote carcinogenesis. Codipilly et al. [26] raised the correlation of increasing obesity in the United States within the 40–59 year old age bracket, and the increased prevalence of Barrett's oesophagus and gastroesophageal reflux disease.

This study demonstrates a significantly improved survival for the >50 years cohort over the last three decades. This is largely due to the improved survival of colorectal cancer alone—a finding noted by Roder et al. [48]. Current oncological treatment targets the carcinogenesis of older-onset gastrointestinal adenocarcinomas, which many investigators globally suggest differ from that of young-onset cancers. Evidence of lower survival rates amongst young-onset cancers in the literature is mixed, with some studies reporting a worse prognosis and limited response to traditional therapy [44], while others are reporting better a prognosis despite more advanced disease [49], and still others indicating similar survival outcomes as compared to their older counterparts [46]. Studies reporting poorer survival rates in the younger population attribute it to more advanced stage disease at the time of diagnosis [26], though this has not been the uniform experience [50]. On the flipside, increased survival rates may be attributed to younger patients having fewer comorbidities and, hence, better responses to chemotherapy and surgery, as well as fewer postoperative complications [46]. Some believe young-onset gastrointestinal cancers to be innately more aggressive and with a differing molecular make up to their older-onset counterparts [5,26,44,50,51]. This study adds to the growing evidence [52,53] of poorer survival amongst males compared to females with young-onset cancers. Sex differences in health outcomes are increasingly being identified and studied. Sex differences in health outcomes have been shown to start in utero. A meta-analysis of RNA sequencing data in fifteen human tissues, including five brain regions, showed differential autosomal and sex chromosome gene expression between males and females in the brain, heart, kidney, colon, and thyroid, and to a lesser extent in bladder, liver, lungs, and pancreas [54]. These may underpin the sexually dimorphic incidence and survival for various cancers.

The rise in incidence of young-onset gastrointestinal adenocarcinomas in South Australia raises more questions and highlights major gaps in our knowledge and understanding of causal mechanisms. It emphasizes the need to explore the environmental and behavioural factors during early life. Consideration of the contribution of perinatal and early-life events in the development of young-onset carcinogenesis is of particular interest [1,13]. While lowering the age for screening programs could result in earlier detection, currently screening in Australia only exists for colorectal cancer for individuals over 50 years [31]. In the case of Barrett's oesophagus, South Australia has a well-established surveillance programme [55,56]. Increased screening carries with it a new set of challenges. The American College of Gastroenterology Guidelines for Colorectal Cancer Screening 2021 [57] recommends the initiation of screening at 45 years instead of 50 years for average-risk individuals, reducing colorectal cancer risk due to earlier detection, and the removal of polyps. It was believed this would reduce the incidence of colorectal cancer in those >50. Yet, a calculated additional 21 million individuals aged 45–49 would need to be screened yearly, creating a significant burden on an already overloaded healthcare system [57]. In South Australia, screening begins at 50 years of age, and patients with a positive faecal immunochemical test (FIT) must undergo colonoscopy within 3 months. Were the screening to drop to 45 years of age, an already strained healthcare system would have to deal with a surge in the requests for colonoscopy with a lack of a clear benefit and would thus present a challenge. As it is unlikely to be a viable or long-term solution, it is imperative we address the factors underlying the rising incidence of young-onset cancers.

One of the major strengths of this study has been the strict inclusion criteria that enabled us to select only oesophageal, stomach, pancreatic, and colorectal adenocarcinoma patients in the past three decades. This may explain the very clear trajectory noted in the studied cancers compared to previous papers from Australia that included all types of cancers affecting a subsite [58,59]. Another strength of this study lies in the appreciation that analysing annual or even four-yearly (Supplementary Figure S1 and Supplementary Table S1) trends in the incidence and survival of diseases, such as cancer, that occur at a lower incidence in smaller populations is fraught with the risk of overlooking important variations over time. The decision to analyse the data in longer time cohorts enabled us to demonstrate the true magnitude of the problem in our region. The data are derived from the South Australian Cancer Registry, a long-standing cancer registry where pathology, death, and clinical reporting is mandatory by law. The advantage of this is rigorous processes of collection of high-quality data. However, it lacks the details of treatment and the stage of the disease at diagnosis. Moreover, the Registry does not contain detailed socio-economic data. These data (on disease stage and sociodemographic variables of race, education, income, etc.) are invaluable to determine the factors involved in the causation of this emerging problem of the rising incidence of young-onset cancers. Being able to decipher the underlying factors will not only reveal any disparities in trends based on demographics, but it will also inform us of the strategies to be employed to prevent the development of these cancers. Nonetheless, this data is invaluable in providing a real-world analysis of young-onset gastrointestinal adenocarcinomas in South Australia.

5. Conclusions

This study from South Australia demonstrates a significant increase in young-onset gastrointestinal adenocarcinomas over the last 28 years, with a greater increase in the male sex. The only significant improvement in survival in this cohort has been noted in colorectal cancer patients. These results signal the need for a concerted global effort in deciphering the causes that underpin the development of young-onset adenocarcinomas, offering the hope of improving outcomes in young cancer patients.

Supplementary Materials: The following supporting information can be downloaded at: https://www.mdpi.com/article/10.3390/cancers14020275/s1. Table S1: Average annual percentage change, AAPC, (Poisson Regression Model) for gender and primary sites of cancer between two age groups (N = 28,566); Table S2: Incidence rate ratios (IRR) and 95% CI (Poisson regression model) for sex and era by primary sites between two age groups (n = 27,855); Table S3: Hazard ratios (HR) and 95% CI (Cox Proportional hazard model) for sex and era by primary sites between two age groups (n = 27,855). Supplementary Figure S1: Trend in incidence rates by sex and era between two age groups across cancer sites 1990–2017 (n = 28,566); Supplementary Figure S2: Kaplan-Meier survival curves for sex, era and primary sites between age groups.

Author Contributions: D.S.: literature search, drafting the manuscript, and final approval; S.U.: design of the study, data analysis, drafting the manuscript, and final approval; M.E.B.-S.: critical revision of manuscript and final approval; M.Y.: critical revision of manuscript and final approval; D.I.W.: critical revision of manuscript and final approval; C.S.K.: critical revision of manuscript and final approval; P.H.: critical revision of manuscript and final approval; S.J.P.: design of the study, critical revision of manuscript, and final approval; C.T.R.: design of the study, critical revision of manuscript, and final approval; S.G.B.: conceptualisation and design, literature search, drafting the manuscript, and final approval. All authors have read and agreed to the published version of the manuscript.

Funding: Our research was supported as follows: SP support—US NIH: R01 AA024464, P01 DK098108, P50 AA0119991, and U01 DK108314; US DoD: W81XWH1910888; CTR support—National Health and Medical Research Council of Australia (NHMRC) Investigator Grant (GNT1174971) and a Matthew Flinders Professorial Fellowship from Flinders University, South Australia.

Institutional Review Board Statement: The study was conducted according to the guidelines of the Declaration of Helsinki, and approved by South Australian Department for Health and Wellbeing Human Research Ethics Committee (HREC) (reference number: LNR/21/SAC/51).

Informed Consent Statement: Patient consent was waived by the Ethics Committee owing to non-identifiable patient data being provided by the South Australian Cancer Registry to the Investigating team.

Data Availability Statement: Authors are unable to provide this data owing to the Ethics approval being granted on the premise that the (South Australian Cancer Registry) data will not be released to a third party.

Conflicts of Interest: The authors declare no conflict of interest.

References

1. Barreto, S.G. We Asked the Experts: Providing the Road Map to Uncovering the Pathophysiology of Young-Onset Cancer to Guide Treatment and Preventive Strategies. *World J. Surg.* **2020**, *44*, 3212–3213. [CrossRef]
2. McWilliams, R.R.; Maisonneuve, P.; Bamlet, W.R.; Petersen, G.M.; Li, D.; Risch, H.A.; Yu, H.; Fontham, E.T.; Luckett, B.; Bosetti, C.; et al. Risk Factors for Early-Onset and Very-Early-Onset Pancreatic Adenocarcinoma: A Pancreatic Cancer Case-Control Consortium (Panc4) Analysis. *Pancreas* **2016**, *45*, 311–316. [CrossRef] [PubMed]
3. Chelmow, D.; Pearlman, M.D.; Young, A.; Bozzuto, L.; Dayaratna, S.; Jeudy, M.; Kremer, M.E.; Scott, D.M.; O'Hara, J.S. Executive Summary of the Early-Onset Breast Cancer Evidence Review Conference. *Obs. Gynecol.* **2020**, *135*, 1457–1478. [CrossRef]
4. Bernards, S.S.; Norquist, B.M.; Harrell, M.I.; Agnew, K.J.; Lee, M.K.; Walsh, T.; Swisher, E.M. Genetic Characterization of Early Onset Ovarian Carcinoma. *Gynecol. Oncol.* **2016**, *140*, 221–225. [CrossRef] [PubMed]
5. van Nistelrooij, A.M.; van Marion, R.; Group, P.; Biermann, K.; Spaander, M.C.; van Lanschot, J.J.; Wijnhoven, B.P.; Dinjens, W.N. Early Onset Esophageal Adenocarcinoma: A Distinct Molecular Entity? *Oncoscience* **2016**, *3*, 42–48. [CrossRef] [PubMed]
6. Setia, N.; Wang, C.X.; Lager, A.; Maron, S.; Shroff, S.; Arndt, N.; Peterson, B.; Kupfer, S.S.; Ma, C.; Misdraji, J.; et al. Morphologic and Molecular Analysis of Early-Onset Gastric Cancer. *Cancer* **2020**, *127*, 103–114. [CrossRef] [PubMed]
7. Murphy, C.C.; Singal, A.G.; Baron, J.A.; Sandler, R.S. Decrease in Incidence of Young-Onset Colorectal Cancer before Recent Increase. *Gastroenterology* **2018**, *155*, 1716–1719.e4. [CrossRef] [PubMed]
8. Rea, D.; Coppola, G.; Palma, G.; Barbieri, A.; Luciano, A.; del Prete, P.; Rossetti, S.; Berretta, M.; Facchini, G.; Perdona, S.; et al. Microbiota Effects on Cancer: From Risks to Therapies. *Oncotarget* **2018**, *9*, 17915–17927. [CrossRef] [PubMed]
9. Farrington, S.M.; McKinley, A.J.; Carothers, A.D.; Cunningham, C.; Bubb, V.J.; Sharp, L.; Wyllie, A.H.; Dunlop, M.G. Evidence for an Age-Related Influence of Microsatellite Instability on Colorectal Cancer Survival. *Int. J. Cancer* **2002**, *98*, 844–850. [CrossRef]
10. Siegel, R.L.; Fedewa, S.A.; Anderson, W.F.; Miller, K.D.; Ma, J.; Rosenberg, P.S.; Jemal, A. Colorectal Cancer Incidence Patterns in the United States, 1974–2013. *J. Natl. Cancer Inst.* **2017**, *109*, djw322. [CrossRef]
11. French, S.A.; Story, M.; Jeffery, R.W. Environmental Influences on Eating and Physical Activity. *Annu. Rev. Public Health* **2001**, *22*, 309–335. [CrossRef]
12. Lui, R.N.; Tsoi, K.K.F.; Ho, J.M.W.; Lo, C.M.; Chan, F.C.H.; Kyaw, M.H.; Sung, J.J.Y. Global Increasing Incidence of Young-Onset Colorectal Cancer across 5 Continents: A Joinpoint Regression Analysis of 1,922,167 Cases. *Cancer Epidemiol. Biomark. Prev.* **2019**, *28*, 1275–1282. [CrossRef]
13. Barreto, S.G.; Pandol, S.J. Young-Onset Carcinogenesis—The Potential Impact of Perinatal and Early Life Metabolic Influences on the Epigenome. *Front. Oncol.* **2021**, *11*, 653289. [CrossRef] [PubMed]
14. Barker, D.J. The Origins of the Developmental Origins Theory. *J. Intern. Med.* **2007**, *261*, 412–417. [CrossRef] [PubMed]
15. Knudson, A.G., Jr. Mutation and Cancer: Statistical Study of Retinoblastoma. *Proc. Natl. Acad. Sci. USA* **1971**, *68*, 820–823. [CrossRef] [PubMed]
16. Lahouel, K.; Younes, L.; Danilova, L.; Giardiello, F.M.; Hruban, R.H.; Groopman, J.; Kinzler, K.W.; Vogelstein, B.; Geman, D.; Tomasetti, C. Revisiting the Tumorigenesis Timeline with a Data-Driven Generative Model. *Proc. Natl. Acad. Sci. USA* **2020**, *117*, 857–864. [CrossRef]
17. Koczwara, B.; Deckx, L.; Ullah, S.; van den Akker, M. Impact of Comorbidities on Physical Function and Survival of Middle-Aged, as Compared to Older, Individuals with Cancer. *Support. Care Cancer* **2021**, *30*, 1625–1632. [CrossRef]
18. Schell, D.; Pandol, S.; Barreto, S.G. Is This the Evidence for Perinatal and Early Life Events Impacting on Cancer Outcomes? *Support. Care Cancer* **2021**. [CrossRef]
19. South Australian Legislation. Health Care Regulations 2008 under the Health Care Act 2008. Available online: https://www.legislation.sa.gov.au/LZ/C/R/Health%20Care%20Regulations%202008.aspx (accessed on 23 November 2021).
20. South Australian Cancer Registry. *Cancer Indicence and Mortality Statistics*; Prevention and Population Health Directorate, Wellbeing SA, Government of South Australia: Adelaide, Australia, 2021.
21. Meng, R.; Chen, J.; D'Onise, K.; Barreto, S.G. Pancreatic Ductal Adenocarcinoma Survival in South Australia: Time Trends and Impact of Tumour Location. *ANZ J. Surg.* **2021**, *91*, 921–926. [CrossRef]
22. Saad El Din, K.; Loree, J.M.; Sayre, E.C.; Gill, S.; Brown, C.J.; Dau, H.; de Vera, M.A. Trends in the Epidemiology of Young-Onset Colorectal Cancer: A Worldwide Systematic Review. *BMC Cancer* **2020**, *20*, 288. [CrossRef]

23. O'Sullivan, D.E.; Hilsden, R.J.; Ruan, Y.; Forbes, N.; Heitman, S.J.; Brenner, D.R. The Incidence of Young-Onset Colorectal Cancer in Canada Continues to Increase. *Cancer Epidemiol.* **2020**, *69*, 101828. [CrossRef]
24. Feletto, E.; Yu, X.Q.; Lew, J.B.; John, D.J.B.S.; Jenkins, M.A.; Macrae, F.A.; Mahady, S.E.; Canfell, K. Trends in Colon and Rectal Cancer Incidence in Australia from 1982 to 2014: Analysis of Data on over 375,000 Cases. *Cancer Epidemiol. Biomark. Prev.* **2019**, *28*, 83–90. [CrossRef]
25. Young, J.P.; Win, A.K.; Rosty, C.; Flight, I.; Roder, D.; Young, G.P.; Frank, O.; Suthers, G.K.; Hewett, P.J.; Ruszkiewicz, A.; et al. Rising Incidence of Early-Onset Colorectal Cancer in Australia over Two Decades: Report and Review. *J. Gastroenterol. Hepatol.* **2015**, *30*, 6–13. [CrossRef]
26. Codipilly, D.C.; Sawas, T.; Dhaliwal, L.; Johnson, M.L.; Lansing, R.; Wang, K.K.; Leggett, C.L.; Katzka, D.A.; Iyer, P.G. Epidemiology and Outcomes of Young-Onset Esophageal Adenocarcinoma: An Analysis from a Population-Based Database. *Cancer Epidemiol. Biomark. Prev.* **2021**, *30*, 142–149. [CrossRef]
27. Sung, J.J.Y.; Chiu, H.M.; Jung, K.W.; Jun, J.K.; Sekiguchi, M.; Matsuda, T.; Kyaw, M.H. Increasing Trend in Young-Onset Colorectal Cancer in Asia: More Cancers in Men and More Rectal Cancers. *Am. J. Gastroenterol.* **2019**, *114*, 322–329. [CrossRef] [PubMed]
28. Lu, X.Q.; Li, Y.; Wang, W.; Feng, W.T.; Shi, O.M.; Wang, Q. International Incidence Trends in Early- and Late-Onset Colorectal Cancer: A Population-Based Study. *Int. J. Colorectal. Dis.* **2020**, *35*, 1077–1086. [CrossRef] [PubMed]
29. Mikaeel, R.R.; Symonds, E.L.; Kimber, J.; Smith, E.; Horsnell, M.; Uylaki, W.; Rico, G.T.; Hewett, P.J.; Yong, J.; Tonkin, D.; et al. Young-Onset Colorectal Cancer Is Associated with a Personal History of Type 2 Diabetes. *Asia Pac. J. Clin. Oncol.* **2021**, *17*, 131–138. [CrossRef] [PubMed]
30. Zhang, J.; Haines, C.; Watson, A.J.M.; Hart, A.R.; Platt, M.J.; Pardoll, D.M.; Cosgrove, S.E.; Gebo, K.A.; Sears, C.L. Oral Antibiotic Use and Risk of Colorectal Cancer in the United Kingdom, 1989–2012: A Matched Case-Control Study. *Gut* **2019**, *68*, 1971–1978. [CrossRef] [PubMed]
31. Lew, J.B.; John, D.J.B.S.; Xu, X.M.; Greuter, M.J.E.; Caruana, M.; Cenin, D.R.; He, E.; Saville, M.; Grogan, P.; Coupe, V.M.H.; et al. Long-Term Evaluation of Benefits, Harms, and Cost-Effectiveness of the National Bowel Cancer Screening Program in Australia: A Modelling Study. *Lancet Public Health* **2017**, *2*, e331–e340. [CrossRef]
32. Gausman, V.; Dornblaser, D.; Anand, S.; Hayes, R.B.; O'Connell, K.; Du, M.; Liang, P.S. Risk Factors Associated with Early-Onset Colorectal Cancer. *Clin. Gastroenterol. Hepatol.* **2020**, *18*, 2752–2759.e2. [CrossRef]
33. Moore, A.; Hikri, E.; Goshen-Lago, T.; Barkan, T.; Morgenstern, S.; Brook, E.; Maderer, A.; Roth, W.; Gordon, N.; Kashtan, H.; et al. Young-Onset Gastric Cancer and Epstein-Barr Virus (Ebv)—A Major Player in the Pathogenesis? *BMC Cancer* **2020**, *20*, 34. [CrossRef]
34. Xie, S.H.; Lagergren, J. The Male Predominance in Esophageal Adenocarcinoma. *Clin. Gastroenterol. Hepatol.* **2016**, *14*, 338–347.e1. [CrossRef] [PubMed]
35. Neufeld, D.; Shpitz, B.; Bugaev, N.; Grankin, M.; Bernheim, J.; Klein, E.; Ziv, Y. Young-Age Onset of Colorectal Cancer in Israel. *Tech. Coloproctol.* **2009**, *13*, 201–204. [CrossRef] [PubMed]
36. Raimondi, S.; Maisonneuve, P.; Lohr, J.M.; Lowenfels, A.B. Early Onset Pancreatic Cancer: Evidence of a Major Role for Smoking and Genetic Factors. *Cancer Epidemiol. Biomark. Prev.* **2007**, *16*, 1894–1897. [CrossRef]
37. Health Policy Centre. *Key Smoking Statistics for SA—2019 Adelaide, Australia*; South Australian Health and Medical Research Institute (SAHMRI): Adelaide, Australia, 2020; Available online: https://sahmri.blob.core.windows.net/communications/sahmri.org/Key_Smoking_Statistics_for_SA_2019_-March_2020.pdf (accessed on 23 November 2021).
38. Mitchell, H.; Katelaris, P. Epidemiology, Clinical Impacts and Current Clinical Management of Helicobacter Pylori Infection. *Med. J. Aust.* **2016**, *204*, 376–380. [CrossRef]
39. Moujaber, T.; MacIntyre, C.R.; Backhouse, J.; Gidding, H.; Quinn, H.; Gilbert, G.L. The Seroepidemiology of Helicobacter Pylori Infection in Australia. *Int. J. Infect. Dis.* **2008**, *12*, 500–504. [CrossRef] [PubMed]
40. O'Connell, J.B.; Maggard, M.A.; Livingston, E.H.; Yo, C.K. Colorectal Cancer in the Young. *Am. J. Surg.* **2004**, *187*, 343–348. [CrossRef]
41. Dozois, E.J.; Boardman, L.A.; Suwanthanma, W.; Limburg, P.J.; Cima, R.R.; Bakken, J.L.; Vierkant, R.A.; Aakre, J.A.; Larson, D.W. Young-Onset Colorectal Cancer in Patients with No Known Genetic Predisposition: Can We Increase Early Recognition and Improve Outcome? *Medicine* **2008**, *87*, 259–263. [CrossRef]
42. Lee, J.; Kim, I.H.; Kim, J.S.; Kim, S.W.; Kim, J.G.; Oh, S.T.; Kang, W.K.; Lee, M.A. Different Clinical Characteristics in Sporadic Young-Age Onset Colorectal Cancer. *Medicine* **2016**, *95*, e4840. [CrossRef]
43. Saluja, S.S.; Manipadam, J.M.; Mishra, P.K.; Sachdeva, S.; Solanki, N.; Shah, H. Young Onset Colorectal Cancer: How Does It Differ from Its Older Counterpart? *Indian J. Cancer* **2014**, *51*, 565–569.
44. Bergquist, J.R.; Leiting, J.L.; Habermann, E.B.; Cleary, S.P.; Kendrick, M.L.; Smoot, R.L.; Nagorney, D.M.; Truty, M.J.; Grotz, T.E. Early-Onset Gastric Cancer Is a Distinct Disease with Worrisome Trends and Oncogenic Features. *Surgery* **2019**, *166*, 547–555. [CrossRef]
45. Piciucchi, M.; Capurso, G.; Valente, R.; Larghi, A.; Archibugi, L.; Signoretti, M.; Stigliano, S.; Zerboni, G.; Barucca, V.; la Torre, M.; et al. Early Onset Pancreatic Cancer: Risk Factors, Presentation and Outcome. *Pancreatology* **2015**, *15*, 151–155. [CrossRef] [PubMed]
46. Ntala, C.; Debernardi, S.; Feakins, R.M.; Crnogorac-Jurcevic, T. Demographic, Clinical, and Pathological Features of Early Onset Pancreatic Cancer Patients. *BMC Gastroenterol.* **2018**, *18*, 139. [CrossRef] [PubMed]

47. Kendall, B.J.; Macdonald, G.A.; Hayward, N.K.; Prins, J.B.; Brown, I.; Walker, N.; Pandeya, N.; Green, A.C.; Webb, P.M.; Whiteman, D.C.; et al. Leptin and the Risk of Barrett's Oesophagus. *Gut* **2008**, *57*, 448–454. [CrossRef]
48. Roder, D.; Karapetis, C.S.; Wattchow, D.; Moore, J.; Singhal, N.; Joshi, R.; Keefe, D.; Fusco, K.; Powell, K.; Eckert, M.; et al. Colorectal Cancer Treatment and Survival: The Experience of Major Public Hospitals in South Australia over Three Decades. *Asia Pac. J. Cancer Prev.* **2015**, *16*, 2431–2440. [CrossRef]
49. Chen, J.N.; Zhang, Q.W.; Pan, Y.B.; Wang, Q.W.; Zhang, X.T.; Li, X.B. Young-Onset Early Colorectal Cancer Had Similar Relative Survival to but Better Overall Survival Than Conventional Early Colorectal Cancer: A Large Population-Based Study. *Front. Oncol.* **2020**, *10*, 96. [CrossRef] [PubMed]
50. Barreto, S.G.; Chaubal, G.N.; Talole, S.; Desouza, A.; Suradkar, K.; Gaikwad, V.; Goel, M.; Shrikhande, S.V. Rectal Cancer in Young Indians-Are These Cancers Different Compared to Their Older Counterparts? *Indian J. Gastroenterol.* **2014**, *33*, 146–150. [CrossRef] [PubMed]
51. Raman, R.; Kotapalli, V.; Adduri, R.; Gowrishankar, S.; Bashyam, L.; Chaudhary, A.; Vamsy, M.; Patnaik, S.; Srinivasulu, M.; Sastry, R.; et al. Evidence for Possible Non-Canonical Pathway(S) Driven Early-Onset Colorectal Cancer in India. *Mol. Carcinog.* **2012**, *53*, E181–E186. [CrossRef] [PubMed]
52. Teng, A.; Lee, D.Y.; Cai, J.; Patel, S.S.; Bilchik, A.J.; Goldfarb, M.R. Patterns and Outcomes of Colorectal Cancer in Adolescents and Young Adults. *J. Surg. Res.* **2016**, *205*, 19–27. [CrossRef]
53. Holowatyj, A.N.; Ruterbusch, J.J.; Rozek, L.S.; Cote, M.L.; Stoffel, E.M. Racial/Ethnic Disparities in Survival among Patients with Young-Onset Colorectal Cancer. *J. Clin. Oncol.* **2016**, *34*, 2148–2156. [CrossRef]
54. Mayne, B.T.; Bianco-Miotto, T.; Buckberry, S.; Breen, J.; Clifton, V.; Shoubridge, C.; Roberts, C.T. Large Scale Gene Expression Meta-Analysis Reveals Tissue-Specific, Sex-Biased Gene Expression in Humans. *Front. Genet.* **2016**, *7*, 183. [CrossRef]
55. Bampton, P.A.; Schloithe, A.; Bull, J.; Fraser, R.J.; Padbury, R.T.; Watson, D.I. Improving Surveillance for Barrett's Oesophagus. *BMJ* **2006**, *332*, 1320–1323. [CrossRef] [PubMed]
56. Vissapragada, R.; Bulamu, N.B.; Brumfitt, C.; Karnon, J.; Yazbeck, R.; Watson, D.I. Improving Cost-Effectiveness of Endoscopic Surveillance for Barrett's Esophagus by Reducing Low-Value Care: A Review of Economic Evaluations. *Surg. Endosc.* **2021**, *35*, 5905–5917. [CrossRef]
57. Shaukat, A.; Kahi, C.J.; Burke, C.A.; Rabeneck, L.; Sauer, B.G.; Rex, D.K. ACG Clinical Guidelines: Colorectal Cancer Screening 2021. *Am. J. Gastroenterol.* **2021**, *116*, 458–479. [CrossRef] [PubMed]
58. Australian Institute of, Health, and Welfare. Cancer Survival and Prevalence in Australia: Period Estimates from 1982 to 2010. *Asia Pac. J. Clin. Oncol.* **2013**, *9*, 29–39. [CrossRef]
59. Cocker, F.; Yee, K.C.; Palmer, A.J.; de Graaff, B. Increasing Incidence and Mortality Related to Liver Cancer in Australia: Time to Turn the Tide. *Aust. N. Z. J. Public Health* **2019**, *43*, 267–273. [CrossRef] [PubMed]

cancers

MDPI

Article

Young-Onset Gastrointestinal Adenocarcinoma Incidence and Survival Trends in the Northern Territory, Australia, with Emphasis on Indigenous Peoples

Mia Shepherdson [1], Shalem Leemaqz [2,3], Gurmeet Singh [4], Courtney Ryder [2,3,5], Shahid Ullah [2], Karla Canuto [6], Joanne P. Young [7,8,9], Timothy J. Price [7,8,9], Ross A. McKinnon [2,3], Stephen J. Pandol [10], Claire T. Roberts [2,3,*] and Savio George Barreto [1,2,3,*]

1 Flinders Medical Center, Bedford Park, Adelaide, SA 5042, Australia; mia.shepherdson@sa.gov.au
2 College of Medicine and Public Health, Flinders University, Adelaide, SA 5001, Australia; shalem.leemaqz@flinders.edu.au (S.L.); courtney.ryder@flinders.edu.au (C.R.); shahid.ullah@flinders.edu.au (S.U.); ross.mckinnon@flinders.edu.au (R.A.M.)
3 Flinders Health and Medical Research Institute, Adelaide, SA 5042, Australia
4 Menzies School of Health Research, Darwin, NT 0810, Australia; gurmeet.singh@menzies.edu.au
5 The George Institute for Global Health, Newtown, NSW 2042, Australia
6 Rural and Remote Health, College of Medicine and Public Health, Flinders University, Darwin, NT 0810, Australia; karla.canuto@flinders.edu.au
7 Department of Haematology and Oncology, The Queen Elizabeth Hospital, Woodville South, SA 5011, Australia; joanne.young@adelaide.edu.au (J.P.Y.); timothy.price@sa.gov.au (T.J.P.)
8 SAHMRI Colorectal Node, Basil Hetzel Institute, Woodville South, SA 5011, Australia
9 Faculty of Health and Medical Sciences, University of Adelaide, Adelaide, SA 5005, Australia
10 Division of Digestive and Liver Diseases, Cedars-Sinai Medical Center, Los Angeles, CA 90048, USA; stephen.pandol@cshs.org
* Correspondence: claire.roberts@flinders.edu.au (C.T.R.); savio.barreto@flinders.edu.au (S.G.B.); Tel.: +61-8-8204-5511 (C.T.R. & S.G.B.)

Citation: Shepherdson, M.; Leemaqz, S.; Singh, G.; Ryder, C.; Ullah, S.; Canuto, K.; Young, J.P.; Price, T.J.; McKinnon, R.A.; Pandol, S.J.; et al. Young-Onset Gastrointestinal Adenocarcinoma Incidence and Survival Trends in the Northern Territory, Australia, with Emphasis on Indigenous Peoples. *Cancers* **2022**, *14*, 2870. https://doi.org/10.3390/cancers14122870

Academic Editor: David Wong

Received: 21 April 2022
Accepted: 6 June 2022
Published: 10 June 2022

Simple Summary: This study from the Australian Northern Territory's Cancer Registry data provides evidence for a significant decrease in incidence of gastrointestinal (oesophageal, stomach, small intestine, colon, rectum, and pancreas) adenocarcinomas over the last 3 decades in individuals aged >50 years, whilst the younger (18–50 years) cohort has remained unchanged with a (non-significant) trend towards an increase. There has been a significantly improved overall survival in both age cohorts. An insight into these trends amongst Australia's Indigenous (who constitute 31% of the territory's population) confirms that while the incidence was significantly lower in Indigenous patients compared to non-Indigenous patients, in both age cohorts, Indigenous patients had worse survival rates. This study calls for a concerted effort aimed at investigating the existence of modifiable sociodemographic factors underlying these disturbing trends. There is a need to enhance preventative strategies, as well as to improve the delivery of cancer care and its uptake amongst Indigenous peoples.

Abstract: Background and Aims: A concerning rise in incidence of young-onset cancers globally led to the examination of trends in incidence and survival of gastrointestinal (GI) adenocarcinomas in the Northern Territory (NT), Australia, over a 28-year period, with a special emphasis on Indigenous peoples. Methods: This cross-sectional analysis of a prospective longitudinal database, NT Cancer Registry (1990–2017), includes all reported cases of GI (oesophagus, gastric, small intestinal, pancreas, colon, and rectum) adenocarcinomas. Poisson regression was used to estimate incidence ratio ratios, and survival was modelled using Cox proportional hazard models separately for people aged 18–50 years and >50 years. Results: A total of 1608 cases of GI adenocarcinoma were recorded during the time of the study. While the overall incidence in people 18–50 years remained unchanged over this time ($p = 0.51$), the rate in individuals aged >50 years decreased (IRR = 0.65 (95% CI 0.56–0.75; $p < 0.0001$)). Incidence rates were significantly less in females >50 years (IRR = 0.67 95% CI 0.59–0.75; $p < 0.0001$), and their survival was significantly better (HR = 0.84 (95%CI 0.72–0.98; $p < 0.03$)) compared to males. Overall survival across all GI subsites improved in both age cohorts, especially between 2010 and 2017 (HR = 0.45 (95%CI 0.29–0.72; $p < 0.0007$) and HR = 0.64 (95%CI 0.52–0.78; $p < 0.0001$), respectively) compared to 1990–1999, driven by an improvement in survival in colonic adenocarcinoma alone, as the survival remained unchanged in other GI subsites. The incidence was significantly lower in Indigenous patients compared to non-Indigenous patients, in both age cohorts (18–50 years IRR = 0.68 95% CI 0.51–0.91; $p < 0.009$ and >50 years IRR = 0.48 95% CI 0.40–0.57; $p < 0.0001$). However, Indigenous patients had worse survival rates (18–50 years HR = 2.06 95% CI 1.36–3.11; $p < 0.0007$ and >50 years HR = 1.66 95% CI 1.32–2.08; $p < 0.0001$). Conclusions: There is a trend towards an increased incidence of young-onset GI adenocarcinomas in the NT. Young Indigenous patients have lower incidence but worse survival across all GI subsites, highlighting significant health inequities in life expectancy. Targeted, culturally safe Indigenous community-focussed programs are needed for early detection and patient-centred management of GI adenocarcinomas.

Keywords: outcomes; morbidity; mortality; stomach; pancreas; colon; Indigenous

1. Introduction

Globally, there have been several reports of increasing incidence of early- or young-onset cancers [1–4]. Recently, we confirmed this disturbing trend over the past 28 years in South Australia by interrogating data from the South Australian Cancer Registry [5]. Early life events have long been suspected to play a role in the causation of young-onset cancers [6,7]. We proposed the Perinatal and Early Life Influences on CANcer (PELICan) hypothesis [8], providing supporting evidence from the literature, linking perinatal stressors that not only affect the epigenome, but also have an increased propensity to affect these children in their adolescent years, resulting in an increased risk of cancer. This can be further influenced by social determinants of health, which produce health inequities between populations. For example, higher cancer incidence rates have been observed in individuals living with a socioeconomic disadvantage [9,10].

Internationally, Indigenous peoples are the longest surviving civilisations on earth and continue to face significant health disparities and marginalisation across multiple social indicators (socioeconomic status, life expectancy, education) from ongoing colonisation [11–15]. In Australia, cancer inequities in outcomes are evident between Indigenous (Aboriginal and Torres Strait Islander) and non-Indigenous patients [16–20]. These disparities prompted the implementation of the National Aboriginal and Torres Strait Islander Cancer Framework in 2014 [21,22]. Given the compelling relationship between young-onset cancers and significant health inequities such as (lower) socioeconomic status, one of our aims was to explore inequity impacts in the incidence and survival of GI adenocarcinomas for Indigenous peoples in the NT. Indigenous Australians make up 3% of Australia's population [23] but account for 31% of the NT population [24]. Understanding manifestations of inequity in young-onset cancer will provide much-needed insight into impacts on Australia's Indigenous and provide critical understanding surrounding cancer preventative strategies (i.e., early detection), which have been created to improve the health of Indigenous Australians.

2. Materials and Methods

The NT Cancer Registry (NTCR) captures all NT cancer diagnosis and cancer-related deaths in accordance with the requirements of the NT Cancer Registration Act (last updated in 2011) [25], specifically, the reporting of all cancer cases to the registry. This activity includes information from treating physicians where the cause of death or patient demographics are incomplete. The registry provides a clinical epidemiological repository of cancers in the NT, to inform government initiatives, as well as established screening and prevention programs. A cross-sectional analysis, from 1 January 1990 to 31 December 2017, of prospectively collected longitudinal data was undertaken, focussing on all cases of adenocarcinoma: oesophagus, stomach, small intestine, pancreas, colon, and rectum.

Ethics approval for this study was obtained from the Aboriginal Ethics Sub-Committee (AESC) of the Human Research Ethics Committee of the NT Department of Health and Menzies School of Health Research (HREC) (HREC Reference Number: 2021-4043).

2.1. Selection of Cases

Inclusion Criteria

NT residents aged >18 years, with a pathologically confirmed diagnosis of GI adenocarcinoma; ICD-10-AM codes C15 oesophagus, C16 stomach, C17 small intestine, C25 pancreas, C18 colon (excluding C18.1), C19 rectosigmoid junction, C20 rectum. Histology codes for adenocarcinoma: ICD-10-AM 8140/2, 8140/3, 8141/3, 8143/3, 8210/2, 8210/3 and 8230/2.

The study period was categorised into 3 time periods (1990–1999, 2000–2009, 2010–2017) to reflect incidence and survivals of cancers over time.

2.2. Statistical Analysis

Statistical analyses were conducted using R version 4.1.1 (R Foundation for Statistical Computing, Vienna, Austria). Patients' characteristics were expressed as median and interquartile range (IQR) for skewed data. The Mann–Whitney U test was used to explore the significance of differences in patients' age between two groups of patients. Proportions were presented as percentages of the respective denominator and were compared between groups using a standard Chi-square test for association with continuity correction, where appropriate.

The incidence rates were calculated by taking the total number of cases divided by the population at risk. The rates were presented per 100,000 persons over 3 time periods for age groups 18–50 years and >50 years, for each sex, Indigenous status and cancer primary sites. Mid-interval population references for each time period were used as the denominator in the calculation of incidence rates. A Poisson regression model was applied to examine the incidence rates between groups of the above characteristics. The estimates were calculated using the likelihood ratio method and were expressed as incidence rate ratios (IRRs) from the Poisson model. The Poisson regression model was also used to calculate the average annual percentage change.

Survival was measured from the date of cancer diagnosis to the date of death, and individuals were censored at date of loss to follow-up or census date. The census date was assigned on 31 December 2017. The NTCR data are linked to the Births, Deaths, and Marriage data annually. Complete data were obtained on survival days, age of diagnosis, date of death, sex, primary sites, and Indigenous status. Cox proportional hazard models were applied to examine the survival outcomes. Sex, primary sites, Indigenous status, and cohort era were used to explore the risk of death between two cohorts (18–50 years and >50 years). The estimates were calculated using the likelihood ratio method and were expressed as hazard ratios (HRs); the lower the HR, the longer the survival. Proportional hazard assumption was tested by a log–log plot of survival and Schoenfeld Residuals. Survival curves for patient survival were evaluated by standard Kaplan–Meier survival curves, and patient cohorts were compared by log-rank test. The two-sided test was performed for all analyses, 95% confidence intervals were reported, and the level of significance was set at 5%.

3. Results

3.1. Patient Data and Overall Incidence and Survival Trends

A total of 1608 patients were diagnosed with oesophageal, stomach, small intestinal, pancreatic, or colorectal adenocarcinoma in the NT between 1990 and 2017 (298, 18.5% patients aged 18–50 years and 1310, 81.5% patients aged >50 years) (Table 1). Adenocarcinoma of the colon was the most commonly reported subsite involved in both age cohorts (27.2/100,000 for individuals aged 18–50 years and 900.3/100,000 for those aged >50 years).

Table 1. Patient's characteristics, era, and primary sites of cancer between two age groups (n = 1608).

	(18–50 Years) n = 298 (18.5%)		(>50 Years) n = 1310 (81.5%)		
	n	%	*n*	%	*p*-Value
Age (years): median (IQR)	45	(39–49)	64	(58–72)	<0.0001
Sex					0.07
Male	172	57.7	834	63.7	
Female	126	42.3	476	36.3	
Indigenous					<0.0001
Non-Indigenous	240	80.5	1179	90.0	
Indigenous	58	19.5	131	10.0	
Era					0.4
1990–1999	72	24.2	275	21.0	
2000–2009	103	34.6	501	38.2	
2010–2017	123	41.3	534	40.8	
Primary site					0.0006
Large intestine (excl. Appendix)	147	49.3	704	53.7	
Rectum	97	32.6	276	21.1	
Pancreas	17	5.7	122	9.3	
Stomach	19	6.4	108	8.2	
Oesophagus	12	4.0	81	6.2	
Small intestine	6	2.0	19	1.5	

Note. Number and percentages are reported unless stated otherwise; IQR, interquartile range. *P*-values are based on Mann–Whitney U test for medians and Chi-square test for proportions.

Age, in itself, was a contributing factor for incidence, with a higher increment per year of age for individuals aged 18–50 years (IRR = 1.15 (95%CI 1.13–1.17; $p < 0.0001$)) as compared to those >50 years (IRR = 1.06 (95%CI 1.06–1.07; $p < 0.0001$)) (Table 2). The overall cancer incidence rates varied by sex for the >50 years cohort (131.1/100,000 for females and 196.9/100,000 for males >50 years) with IRR = 0.67 (95%CI 0.59–0.75; $p < 0.0001$). There was a significantly lower incidence of adenocarcinomas in every subsite for females in the >50 years cohort, except for pancreatic adenocarcinoma (Figure 1). In the 18–50 years cohort, no significant difference in incidence, by sex of the patient, was noted either in the overall cohort (4.4/100,000 for females and 5.4/100,000 for males aged 18–50 years; $p = 0.09$), or by organ subsite in the 18–50 years cohort (Table 2, Figures 1 and S1).

There was a significantly improved survival across all GI adenocarcinoma subsites over the 3 time cohorts (HR = 0.45 (95%CI 0.29–0.72; $p = 0.0007$)) for those aged 18–50 years and 0.64 (95%CI 0.52–0.78; $p < 0.0001$) for those >50 years, respectively, between 2010 and 2017 compared to years 1990 to 1999. Here also, females >50 years had a significantly improved survival compared to their male counterparts (HR = 0.84 (95%CI 0.72–0.98; $p = 0.03$)). No significant difference in survival was noted amongst females in the 18–50 years cohort (HR = 0.88 (95%CI 0.62–1.27; $p = 0.5$)).

The improvement in survival amongst the 18–50 years cohort was due only to a significantly improved survival for colonic adenocarcinoma in the last time period (HR = 0.46 (95%CI 0.22–0.98; $p = 0.04$)). On the contrary, in the >50 years cohort, the significantly improved survival was driven by improvements in survival in colonic (HR = 0.60 (95%CI 0.44–0.81; $p = 0.0008$)), oesophageal (HR = 0.41 (95%CI 0.20–0.84; $p = 0.02$)), and small intestinal (HR = 0.11 (95%CI 0.01–0.83; $p = 0.03$)) adenocarcinomas over the last time period (Figure 2).

Table 2. Incidence rates (IR) and incidence rate ratios (IRR) for age, sex, era, and cancer sites between two age groups (n = 1608).

	(18–50 Years)			(>50 Years)		
	n = 298			n = 1310		
	* IR (95% CI)	IRR (95% CI)	p-Value	* IR (95% CI)	IRR (95% CI)	p-Value
Overall	95.64 (85.09, 107.14)			1321.70 (1251.09, 1395.26)		
Age (years)	-	1.15 (1.13, 1.17)	<0.0001	-	1.06 (1.06, 1.07)	<0.0001
Sex						
Male	5.39 (4.09, 7.10)	Reference	-	196.86 (172.34, 224.87)	Reference	-
Female	4.41 (3.31, 5.87)	0.82 (0.65, 1.03)	0.09	131.05 (113.74, 151.01)	0.67 (0.59, 0.75)	<0.0001
Indigenous						
Non-Indigenous	5.91 (4.61, 7.56)	Reference	-	232.99 (209.68, 258.88)	Reference	-
Indigenous	4.02 (2.87, 5.62)	0.68 (0.51, 0.91)	0.009	110.73 (91.44, 134.09)	0.48 (0.40, 0.57)	<0.0001
Era						
1990–1999	4.66 (3.40, 6.38)	Reference	-	189.49 (161.17, 222.79)	Reference	-
2000–2009	4.83 (3.56, 6.57)	1.04 (0.77, 1.41)	0.81	177.91 (154.23, 205.23)	0.94 (0.81, 1.09)	0.40
2010–2017	5.14 (3.82, 6.90)	1.10 (0.82, 1.48)	0.51	122.92 (106.80, 141.48)	0.65 (0.56, 0.75)	<0.0001
Cancer site						
Large intestine	27.22 (21.68, 34.18)	-	-	900.31 (807.56, 1003.71)	-	-
Rectum	17.96 (13.91, 23.19)			351.68 (304.95, 405.58)		
Pancreas	3.15 (1.91, 5.20)	-	-	156.02 (128.44, 189.53)	-	-
Stomach	3.52 (2.18, 5.67)	-	-	138.12 (112.54, 169.50)	-	-
Oesophagus	2.22 (1.23, 4.00)	-	-	103.59 (82.15, 130.63)	-	-
Small intestine	1.11 (0.49, 2.51)	-	-	24.30 (15.39, 38.36)	-	-

* IR is incidence per 100,000 Northern Territory residents. IRs were not reported for age, and IRRs were not reported for cancer sites.

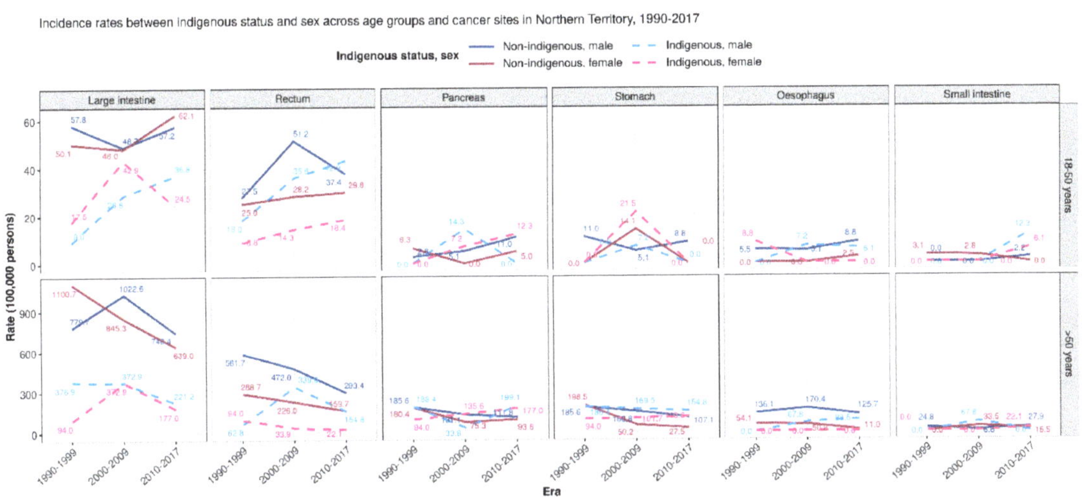

Figure 1. Trend in incidence rates by Indigenous status, sex, and era between two age groups across cancer sites, 1990–2017 (n = 1608).

3.1.1. GI adenocarcinomas Trends across Populations

The overall incidence rate for GI adenocarcinomas in individuals aged 18–50 years in the NT remained unchanged over the last three time periods from 1990–1999 to 2010–2017 (4.7/100,000 to 5.1/100,000), and the rates amongst individuals aged >50 years decreased over the same time period from 189.5/100,000 to 122.9/100,000 (IRR = 0.65 (95%CI 0.56–0.75); $p < 0.0001$) (Table 2). Cancer incidence rates for Indigenous peoples aged 18–50 years were 4.0/100,000 and 110.7/100,000 for >50 years of ages, which was significantly lower compared to non-Indigenous peoples, 5.9/100,000 (18–50 years) with IRR = 0.68 (95%CI 0.51–0.91) ($p = 0.009$) and 233/100,000 (>50 years) with IRR = 0.48 (95%CI 0.40–0.57) ($p < 0.0001$).

Cohort	Sites	Groups	HR (95% CI)	p-value
18-50 years	Large intestine	Male	Reference	
		Female	0.82 (0.47, 1.44)	0.5
		Non-indigenous	Reference	
		Indigenous	2.22 (1.06, 4.64)	0.03
		1990-1999	Reference	
		2000-2009	0.73 (0.38, 1.44)	0.37
		2010-2017	0.46 (0.22, 0.98)	0.04
	Rectum	Male	Reference	
		Female	0.53 (0.25, 1.16)	0.11
		Non-indigenous	Reference	
		Indigenous	2.55 (1.13, 5.74)	0.02
		1990-1999	Reference	
		2000-2009	0.74 (0.33, 1.66)	0.47
		2010-2017	0.47 (0.17, 1.29)	0.15
	Pancreas	Male	Reference	
		Female	1.40 (0.11, 18.24)	0.8
		Non-indigenous	Reference	
		Indigenous	0.25 (0.01, 8.64)	0.45
		1990-1999	Reference	
		2000-2009	0.59 (0.03, 13.03)	0.74
		2010-2017	0.21 (0.02, 2.73)	0.23
	Stomach	Male	Reference	
		Female	1.29 (0.13, 12.57)	0.83
		Non-indigenous	Reference	
		Indigenous	6.30 (0.75, 53.27)	0.09
		1990-1999	Reference	
		2000-2009	0.27 (0.03, 2.77)	0.27
		2010-2017	0.88 (0.16, 4.97)	0.88
	Oesophagus⁺	Male	Reference	
		Female	-	-
		Non-indigenous	Reference	
		Indigenous	-	-
		1990-1999	Reference	
		2000-2009	-	-
		2010-2017	-	-
	Small intestine⁺	Male	Reference	
		Female	-	-
		Non-indigenous	Reference	
		Indigenous	-	-
		1990-1999	Reference	
		2000-2009	-	-
		2010-2017	-	-
>50 years	Large intestine	Male	Reference	
		Female	0.83 (0.67, 1.02)	0.08
		Non-indigenous	Reference	
		Indigenous	1.72 (1.19, 2.48)	0.004
		1990-1999	Reference	
		2000-2009	0.79 (0.61, 1.02)	0.07
		2010-2017	0.60 (0.44, 0.81)	0.0008
	Rectum	Male	Reference	
		Female	0.92 (0.63, 1.35)	0.68
		Non-indigenous	Reference	
		Indigenous	1.37 (0.75, 2.49)	0.31
		1990-1999	Reference	
		2000-2009	0.78 (0.53, 1.14)	0.2
		2010-2017	0.64 (0.39, 1.04)	0.07
	Pancreas	Male	Reference	
		Female	1.10 (0.75, 1.61)	0.64
		Non-indigenous	Reference	
		Indigenous	1.84 (1.17, 2.89)	0.008
		1990-1999	Reference	
		2000-2009	0.89 (0.54, 1.48)	0.66
		2010-2017	0.75 (0.46, 1.21)	0.23
	Stomach	Male	Reference	
		Female	0.95 (0.57, 1.60)	0.86
		Non-indigenous	Reference	
		Indigenous	1.08 (0.62, 1.88)	0.78
		1990-1999	Reference	
		2000-2009	1.19 (0.68, 2.07)	0.54
		2010-2017	0.77 (0.42, 1.40)	0.39
	Oesophagus	Male	Reference	
		Female	0.59 (0.28, 1.23)	0.16
		Non-indigenous	Reference	
		Indigenous	3.85 (1.68, 8.82)	0.001
		1990-1999	Reference	
		2000-2009	0.52 (0.27, 1.01)	0.05
		2010-2017	0.41 (0.20, 0.84)	0.02
	Small intestine	Male	Reference	
		Female	1.49 (0.18, 12.25)	0.71
		Non-indigenous	Reference	
		Indigenous	5.43 (0.17, 178.26)	0.34
		1990-1999	Reference	
		2000-2009	0.06 (0.00, 1.53)	0.09
		2010-2017	0.11 (0.01, 0.83)	0.03

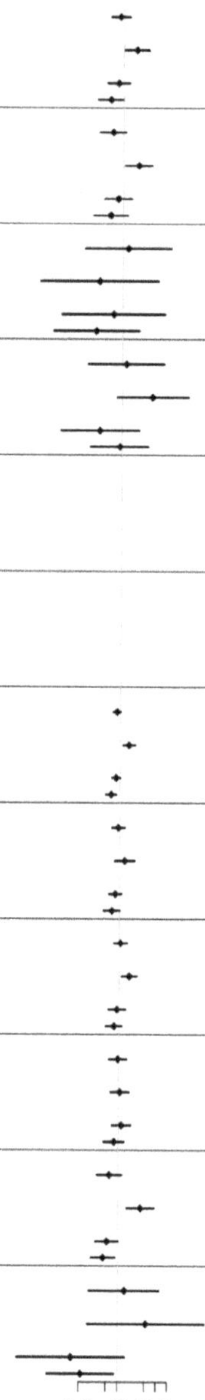

Figure 2. Hazard ratios (HR) and 95% CI (Cox proportional hazard model) for Indigenous status, sex, and era by primary sites between two age groups ($n = 1608$). ⁺ Reliable estimates of HR (95% CI) cannot be obtained for Oesophagus and Small intestine in 18–50 years due to small sample size.

The reduced overall incidence of GI adenocarcinomas noted amongst Indigenous peoples in both age cohorts was largely influenced by the significantly lower incidence of colonic adenocarcinomas in the 18–50 years cohort (IRR = 0.51 (95%CI 0.32–0.79, $p = 0.003$)), and colonic and rectal adenocarcinomas in the >50 years cohort (IRR = 0.31 (95%CI 0.23–0.41, $p < 0.0001$); and (IRR = 0.38 (95%CI 0.25–0.59, $p < 0.0001$), respectively) (Figure 3 and Supplementary Table S2).

Cohort	Sites	Groups	IRR (95% CI)	*p*-value
18-50 years	Large intestine	Male	Reference	
		Female	1.03 (0.74, 1.42)	0.87
		Non-indigenous	Reference	
		Indigenous	0.51 (0.32, 0.79)	0.003
		1990-1999	Reference	
		2000-2009	0.80 (0.52, 1.23)	0.32
		2010-2017	0.95 (0.64, 1.43)	0.82
	Rectum	Male	Reference	
		Female	0.65 (0.43, 0.99)	0.04
		Non-indigenous	Reference	
		Indigenous	0.75 (0.46, 1.24)	0.26
		1990-1999	Reference	
		2000-2009	1.32 (0.77, 2.26)	0.31
		2010-2017	1.24 (0.73, 2.11)	0.43
	Pancreas	Male	Reference	
		Female	0.77 (0.29, 2.01)	0.59
		Non-indigenous	Reference	
		Indigenous	1.22 (0.43, 3.49)	0.71
		1990-1999	Reference	
		2000-2009	1.28 (0.30, 5.41)	0.74
		2010-2017	2.02 (0.54, 7.53)	0.29
	Stomach	Male	Reference	
		Female	0.81 (0.33, 2.02)	0.65
		Non-indigenous	Reference	
		Indigenous	0.78 (0.26, 2.37)	0.66
		1990-1999	Reference	
		2000-2009	1.96 (0.62, 6.25)	0.25
		2010-2017	0.63 (0.16, 2.56)	0.52
	Oesophagus	Male	Reference	
		Female	0.22 (0.05, 1.02)	0.05
		Non-indigenous	Reference	
		Indigenous	0.95 (0.25, 3.55)	0.93
		1990-1999	Reference	
		2000-2009	0.74 (0.15, 3.75)	0.72
		2010-2017	1.31 (0.32, 5.33)	0.70
	Small intestine	Male	Reference	
		Female	1.06 (0.21, 5.27)	0.94
		Non-indigenous	Reference	
		Indigenous	2.90 (0.58, 14.49)	0.20
		1990-1999	Reference	
		2000-2009	0.88 (0.05, 14.29)	0.93
		2010-2017	3.03 (0.34, 27.42)	0.32
>50 years	Large intestine	Male	Reference	
		Female	0.86 (0.74, 1.00)	0.06
		Non-indigenous	Reference	
		Indigenous	0.31 (0.23, 0.41)	<0.0001
		1990-1999	Reference	
		2000-2009	1.07 (0.87, 1.32)	0.52
		2010-2017	0.72 (0.59, 0.89)	0.002
	Rectum	Male	Reference	
		Female	0.45 (0.35, 0.59)	<0.0001
		Non-indigenous	Reference	
		Indigenous	0.38 (0.25, 0.59)	<0.0001
		1990-1999	Reference	
		2000-2009	0.87 (0.64, 1.18)	0.38
		2010-2017	0.52 (0.38, 0.71)	<0.0001
	Pancreas	Male	Reference	
		Female	0.79 (0.55, 1.13)	0.19
		Non-indigenous	Reference	
		Indigenous	1.20 (0.78, 1.84)	0.41
		1990-1999	Reference	
		2000-2009	0.60 (0.37, 0.99)	0.04
		2010-2017	0.67 (0.43, 1.04)	0.08
	Stomach	Male	Reference	
		Female	0.47 (0.31, 0.71)	0.0003
		Non-indigenous	Reference	
		Indigenous	1.28 (0.82, 2.00)	0.28
		1990-1999	Reference	
		2000-2009	0.64 (0.39, 1.02)	0.06
		2010-2017	0.45 (0.28, 0.71)	0.0008
	Oesophagus	Male	Reference	
		Female	0.18 (0.10, 0.34)	<0.0001
		Non-indigenous	Reference	
		Indigenous	0.36 (0.16, 0.84)	0.02
		1990-1999	Reference	
		2000-2009	1.25 (0.67, 2.33)	0.48
		2010-2017	0.78 (0.42, 1.47)	0.45
	Small intestine	Male	Reference	
		Female	0.82 (0.33, 2.06)	0.68
		Non-indigenous	Reference	
		Indigenous	0.82 (0.24, 2.83)	0.76
		1990-1999	Reference	
		2000-2009	1.83 (0.38, 8.81)	0.45
		2010-2017	1.69 (0.37, 7.73)	0.50

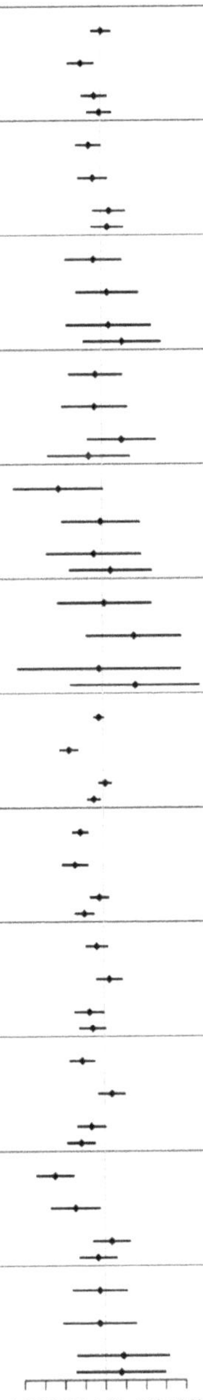

Figure 3. Incidence rate ratios (IRR) and 95% CI (Poisson regression model) for Indigenous status, sex, and era by primary sites between two age groups ($n = 1608$).

Rates of GI adenocarcinomas increased per year in males 18–50 years, with an estimated average annual percentage change (AAPC) of 2.23% (95%CI 0.32–4.18; $p = 0.02$). There was also an increasing trend for both non-Indigenous and Indigenous populations in 18–50 years, with the Indigenous population having a higher AAPC (AAPC 1.88 (95%CI 0.28–3.51; $p = 0.02$) for non-Indigenous and AAPC 3.70 (95%CI 0.24–7.28; $p = 0.04$)) (Supplementary Table S1).

3.1.2. Survival by Time Trends and Site in Indigenous Compared to Non-Indigenous Peoples

Survival estimates from Kaplan–Meier plots revealed the greatest median survivals for rectal adenocarcinoma in 18–50 years (3.98 years) and colonic adenocarcinoma in >50 years (4.17 years) (Table 3). Using colonic adenocarcinoma as the reference, survival following pancreatic and stomach adenocarcinoma were significantly lower for those aged 18–50 years (HR = 2.30 (95%CI 1.19–4.46; $p < 0.01$) and HR = 2.73 (95%CI 1.49–5.02; $p < 0.001$), respectively). For those aged >50 years, survival following pancreatic (HR = 5.76 (95%CI 4.58–7.24; $p < 0.0001$)), stomach (HR = 2.91 (95%CI 2.28–3.70; $p < 0.0001$)), oesophageal (HR = 3.29 (95%CI 2.51–4.30; $p < 0.0001$)), and small intestinal (HR = 2.01 (95%CI 1.10–3.68; $p < 0.02$)) adenocarcinoma was significantly lower compared to the reference (Table 4 and Supplementary Figure S2).

Table 3. Median survival times (years) for primary sites of cancer between two age groups ($n = 1608$).

Cancer Sites	(18–50 Years)	(>50 Years)
	n = 298 (18.5%)	*n* = 1310 (81.5%)
	Median (95% CI)	Median (95% CI)
Overall	3.37 (1.59–8.66)	2.91 (1.24–7.46)
Large intestine	3.75 (1.79–9.97)	4.17 (1.75–8.45)
Rectum	3.98 (1.90–9.62)	4.06 (1.67–9.03)
Pancreas	1.59 (0.97–5.59)	0.98 (0.56–1.49)
Stomach	1.51 (1.12–2.79)	1.55 (0.85–2.81)
Oesophagus	2.19 (1.56–3.56)	1.64 (1.06–2.57)
Small intestine	1.73 (0.80–4.19)	1.86 (0.82–4.98)

Table 4. Hazard ratios (HR) and 95% CI (Cox proportional hazard model) for age, sex, era, and cancer sites between two age groups ($n = 1608$).

	(18–50 Years)		(>50 Years)	
	n = 298		*n* = 1310	
	HR (95% CI)	*p*-Value	HR (95% CI)	*p*-Value
Age (years)	1.02 (1.00, 1.05)	0.07	1.04 (1.03, 1.05)	<0.0001
Sex				
Male	Reference	-	Reference	-
Female	0.88 (0.62, 1.27)	0.5	0.84 (0.72, 0.98)	0.03
Indigenous				
Non-Indigenous	Reference	-	Reference	-
Indigenous	2.06 (1.36, 3.11)	0.0007	1.66 (1.32, 2.08)	<0.0001
Era				
1990–1999	Reference	-	Reference	-
2000–2009	0.63 (0.42, 0.97)	0.03	0.81 (0.68, 0.97)	0.02
2010–2017	0.45 (0.29, 0.72)	0.0007	0.64 (0.52, 0.78)	<0.0001
Cancer site				
Large intestine	Reference		Reference	
Rectum	0.94 (0.63, 1.42)	0.78	1.01 (0.82, 1.23)	0.95
Pancreas	2.30 (1.19, 4.46)	0.01	5.76 (4.58, 7.24)	<0.0001
Stomach	2.73 (1.49, 5.02)	0.001	2.91 (2.28, 3.70)	<0.0001
Oesophagus	1.87 (0.88, 3.98)	0.1	3.29 (2.51, 4.30)	<0.0001
Small intestine	1.88 (0.59, 6.02)	0.29	2.01 (1.10, 3.68)	0.02

Survival, following GI adenocarcinomas, was significantly lower amongst Indigenous peoples as compared to non-Indigenous peoples in both age cohorts over the study period (HR = 2.06 (95%CI 1.36–3.11; $p < 0.0007$) for the 18–50 years cohort and HR = 1.66 (95%CI 1.32–2.08; $p < 0.0001$) for the >50 years cohort, respectively) (Table 4 and Supplementary Figure S2). The significantly reduced survival for Indigenous peoples was largely the effect of colonic and rectal adenocarcinoma in those aged 18–50 years (HR = 2.22 (95%CI 1.06–4.64; $p < 0.03$) and HR = 2.55 (95%CI 1.13–5.74; $p < 0.02$), respectively). For Indigenous people >50 years, reduced survival was largely the effect of colonic (HR = 01.72 (95%CI 1.19–2.48; $p < 0.004$)), pancreatic (HR = 1.84 (95%CI 1.17–2.89; $p < 0.008$)), and oesophageal (HR = 3.85 (95%CI 1.68–8.82; $p < 0.001$)) adenocarcinomas (Supplementary Table S3).

4. Discussion

Our research shares the incidence and survival of GI adenocarcinomas in NT residents over a 28-year period. The incidence of young-onset cancers (18–50 years) marginally, but not statistically significantly, increased (by 10%), while the incidence in the >50 years cohort significantly reduced. Male sex was associated with poorer prognosis in older age (>50 year), whereas young-onset cancers were non-discriminatory. Poorer survival outcomes were observed in Indigenous populations despite an overall lower incidence of GI adenocarcinomas, compared to non-Indigenous individuals. This is a novel finding for young-onset cancers.

Lifestyle and environmental factors such as obesity, diet, smoking, and alcohol exposure are well-documented risk factors for young-onset GI cancers [26–28]. The NT records the highest percentage of both daily smokers (20%) and alcohol consumers who exceed the recommended number of daily standard drinks (21.4%), compared to other states and territories in Australia. The association between rising smoking rates and the rise in tobacco-related cancers was reported in the NT nearly a decade ago [29]. Metabolic syndrome and its contributing factors of poor diet and sedentary lifestyle have been linked to early-onset solid-organ tumours [27,28]. Liu et al. [28] demonstrated a risk ratio of 1.2 for every 5-unit increment in body mass index (BMI) in young adults diagnosed with colorectal cancer. Over the past two decades, we have seen an increase in both obesity and young-onset cancers in the Australian population [30]. This suggests a temporal association of metabolic syndrome serving as a contributory factor to the rise in young-onset cancers. However, it does not explain the stable rate of late-onset cancers despite an increase in BMI in this age group, too. Although the incidence of GI adenocarcinomas is on the rise, improved survival has been demonstrated in our young-onset cohort. This finding is in contrast to our observations in young-onset adenocarcinomas in South Australia (where the overall survival across all subsites has remained unchanged) [5]. This trend has been attributed to the significantly improved survival in colonic adenocarcinoma, especially in the last time period. Improved survival in colonic adenocarcinoma may be credited to increasing awareness of young-onset cancers or timely access to imaging and colonoscopy. Additionally, the cancer surveillance programmes (colon, breast, and cervical) are incorporated as part of primary health care and specialist outreach services in the NT. Colon cancer is also likely to present earlier than stomach and pancreatic cancer, thus allowing for a variety of treatment options with less disease burden [31]. Improved colon cancer survival is unlikely related to screening in patients with a positive family history, as young-onset colorectal cancer is more commonly diagnosed in patients without a family history [32]. The lack of hereditary predisposition has also been noted in young-onset gastric and pancreatic cancers, suggesting young-onset cancers are the result of an alternate carcinogenic pathway [33–35]. For this reason, simply reducing the age of screening based on family history will likely increase the burden on the health care system without improving young-onset cancer survival significantly. However, a recent modelling study found that reducing the age for Aboriginal and Torres Strait Islander peoples would be cost-effective and save more lives [36]. Additionally, investing resources into studying the cause of increased young-onset GI cancers may be more economical and beneficial for society in the long term.

Male gender has been consistently associated with greater incidence and poorer survival in GI cancers [37,38]. The Australian Bureau of Statistics reports that adult males are more likely to be obese and partake in higher rates of smoking and alcohol consumption compared to females [30,39,40]. It is therefore not surprising older males are at higher risk of developing and dying from GI cancers. Perhaps the most pertinent finding when analysing sex in our study was that the young-onset GI cancers did not discriminate between male and female patients. This may be partially explained by the PELICan hypothesis [8] as exposure to perinatal stressors and carcinogens in utero would be equal amongst male and female foetuses. Unborn males would not be at an increased risk of GI cancers compared to females as smoking, obesity, and alcohol are male-dominated risk factors experienced later in life. The PELICan hypothesis may also explain how the incidence of GI cancers was stable for the >50 years cohort but increasing for the 18–50 years cohort. Not only has obesity increased in the adult population, but also in the pregnant population, where currently 1 in 5 pregnant women are obese [41,42]. Obesity in pregnancy with its associated perinatal complications [43] may result in stress and/or inflammation-induced epigenetic changes in the foetus predisposing to obesity and young-onset cancer.

The impact of colonisation globally continues to impact the socioeconomic and health status and life expectancy for Indigenous populations as compared to non-Indigenous populations. Social determinants and lower socioeconomic status have been consistently associated with poor cancer survival [44,45]. In 2006, Anderson et al. [46] alarmingly reported that the life expectancy of Aboriginal and Torres Strait Islander peoples was 20 years lower than that for the total Australian population. The recent statistics from the Australian Institute of Health and Welfare [47] show improvements, but not parity. In 2015–2017, life expectancy at birth for Indigenous Australians was estimated to be 71.6 years for males and 75.6 years for females. In comparison, over the same period, life expectancy at birth for non-Indigenous Australians was 80.2 years for males and 83.4 years for females. Cancer survival was no exception to this health discrepancy, which prompted the Australian government to implement the National Aboriginal and Torres Strait Islander Cancer Framework in 2014 [21,22,48]. The results of our study are consistent with the literature whereby, despite a lower incidence of cancer, Indigenous patients had a significantly reduced survival in both age cohorts. These outcomes concur in part with research by Condon et al. [17], who also demonstrated poorer cancer survival amongst Indigenous peoples in the NT, with diagnoses at an advanced stage of the cancer being a contributing factor. Additionally, a previous study from South Australia has flagged issues around availability and access to surgical and systemic treatments amongst Aboriginal cancer patients compared to non-Aboriginal South Australian patients, which further complicated the disadvantages associated with geographic remoteness and advanced stage of disease at diagnosis, compounded by the presence of associated comorbid conditions [49]. In the present study, owing to the lack of stage-specific data, we are unable to comment on whether a delayed diagnosis was responsible for the poorer survival observed. Certainly, the significant health inequities that impact Indigenous Australians with ongoing marginalisation would be contributors to this trend [16]. Distance and access to tertiary health settings have been shown to contribute to later presentation and more advanced disease in Indigenous populations, along with health professionals in rural and remote settings being trained as generalists [18,50,51]. Not only is remoteness a likely impediment to regular, and timely, access to health care and poor overall cancer survival [52], but cultural marginalisation and access to culturally safe and responsive healthcare is also a compounding factor contributing to delayed disease presentations or delay in timely treatment [16,53]. Whilst underrepresentation of Indigenous status within registries remains an area of global concern [12,20], in the NT, there is good ascertainment of Indigenous status, as evidenced by no missing data in the NTCR data analysed in this study.

This study successfully reported the incidence and survival of GI adenocarcinoma in the NT and evaluated the influence of Indigenous status, age, and primary tumour location. Data were collected from the reputable NTCR. The use of three distinct time periods was effective in tracking changes in incidence and survival over time and allowed for a quick comparison to our previous South Australian study. The study could be further improved by data linkage, including data pertaining to patient socioeconomic status and premorbid baseline such as postcode, BMI, alcohol intake, and smoking status. Data linkage with national administrative and clinical datasets (e.g., Pharmaceutical Benefits Scheme/PBS and Medicare Benefits Schedule/MBS) will help provide pertinent information on costs incurred with the management of these cancers. Information on timely access to cancer treatment such as chemoradiotherapy or surgical management may help explain survival outcomes. Data stratification resulted in small cohort sample sizes, which limited the statistical power and ability to conduct an in-depth analysis on the significance of primary tumour location. Further studies with an increased sample size will likely yield information on the behaviour of different gastrointestinal cancers, which could influence current screening regimes. In order to capture data that appropriately record factors impacting on patient outcomes, data registries should engage with Aboriginal and Torres Strait Islander health bodies to broaden coding, to capture community needs, and work towards data sovereignty. Data that recognise and respond to the health and well-being concepts and needs of Australia's First Peoples constitutes a step towards data sovereignty for Aboriginal and Torres Strait Islander communities [54,55]. An exemplar of this includes the Footprints in Time Study, with Indigenous leadership, oversight, and a focus on positive strength-based data collection and reporting [56].

5. Conclusions

Our study uniquely compares the incidence and survival in young-onset GI adenocarcinomas between Indigenous and non-Indigenous NT residents. This study demonstrates that not only are young-onset GI cancers increasing for residents aged 18–50 years, but there is a significant and disturbing trend of lower incidence but poorer survival for Indigenous residents of any age. Lifestyle and environmental factors during the perinatal period and into early adulthood are likely contributors to these phenomena; however, more research is imperative to identify at-risk cohorts.

Supplementary Materials: The following supporting information can be downloaded at: https://www.mdpi.com/article/10.3390/cancers14122870/s1. Table S1: Average annual percentage change, AAPC, (Poisson Regression Model) for gender, Indigenous status and primary sites of cancer between two age groups (n = 1608); Table S2: Incidence rate ratios (IRR) and 95% CI (Poisson regression model) for sex, Indigenous status and era by primary sites between two age groups (n = 1608); Table S3: Hazard ratios (HR) and 95% CI (Cox Proportional hazard model) for sex, Indigenous status and era by primary sites between two age groups (n = 1608); Supplementary Figure S1: Trend in incidence rates by sex, Indigenous status and era between two age groups across cancer sites 1990–2017 (n = 1608); Figure S2: Kaplan-Meier survival curves for sex, era, Indigenous status and primary sites between age groups.

Author Contributions: Conceptualisation, S.G.B.; Formal analysis, S.L. and G.S.; Investigation, M.S.; Methodology, S.G.B., M.S., S.L., C.R., S.J.P. and C.T.R.; Project administration, S.G.B. and C.R.; Software, S.L.; Supervision, S.G.B. and C.T.R.; Validation, S.U., K.C., J.P.Y., T.J.P., R.A.M. and S.J.P.; Writing—original draft, M.S.; Writing—review & editing, S.G.B., S.L., G.S., C.T.R., S.U., K.C., J.P.Y., T.J.P., R.A.M., S.J.P. and C.T.R. All authors have read and agreed to the published version of the manuscript.

Funding: RAM support: supported by the Beat Cancer Project funded by Cancer Council SA and SA Health. SP support—US NIH: R01 AA024464, P01 DK098108, P50 AA0119991, U01 DK108314, US DoD: W81XWH1910888. CTR support—National Health and Medical Research Council of Australia (NHMRC) Investigator Grant (GNT1174971) and a Matthew Flinders Professorial Fellowship from Flinders University, South Australia. SGB support—Flinders Foundation grant: 49358025.

Institutional Review Board Statement: The study was conducted according to the guidelines of the Declaration of Helsinki. Ethics approval for this study was obtained from the Aboriginal Ethics Sub-Committee (AESC) of the Human Research Ethics Committee of the NT Department of Health and Menzies School of Health Research (HREC) (HREC Reference Number: 2021-4043).

Informed Consent Statement: Since only non-identifiable data was provided to us by the NTCR, a waiver of consent was provided by the Ethics Committee.

Data Availability Statement: Authors are unable to provide this data owing to the Ethics approval being granted on the premise that the (Northern Territory Cancer Registry) data will not be released to a third party.

Acknowledgments: Shu Qin Li, Northern Territory Cancer Registrar, for her administrative support to this project.

Conflicts of Interest: The authors declare no conflict of interest.

References

1. Barreto, S.G.; Pandol, S.J. Young-Onset Carcinogenesis—The Potential Impact of Perinatal and Early Life Metabolic Influences on the Epigenome. *Front. Oncol.* **2021**, *11*, 653289. [CrossRef] [PubMed]
2. Done, J.Z.; Fang, S.H. Young-Onset Colorectal Cancer: A Review. *World J. Gastrointest. Oncol.* **2021**, *13*, 856–866. [CrossRef] [PubMed]
3. El Din, K.S.; Loree, J.; Sayre, E.C.; Gill, S.; Brown, C.; Dau, H.; De Vera, M.A. Trends in the epidemiology of young-onset colorectal cancer: A worldwide systematic review. *BMC Cancer* **2020**, *20*, 288. [CrossRef]
4. di Martino, E.; Smith, L.; Bradley, S.H.; Hemphill, S.; Wright, J.; Renzi, C.; Bergin, R.; Emery, J.; Neal, R.D. Incidence trends for twelve cancers in younger adults—A rapid review. *Br. J. Cancer* **2022**, *126*, 1374–1386. [CrossRef] [PubMed]
5. Schell, D.; Ullah, S.; Brooke-Smith, M.E.; Hollington, P.; Yeow, M.; Karapetis, C.S.; Watson, D.I.; Pandol, S.; Roberts, C.T.; Barreto, S.G. Gastrointestinal Adenocarcinoma Incidence and Survival Trends in South Australia, 1990–2017. *Cancers* **2022**, *14*, 275. [CrossRef] [PubMed]
6. Murphy, C.C.; Singal, A.G.; Baron, J.A.; Sandler, R.S. Decrease in Incidence of Young-Onset Colorectal Cancer Before Recent Increase. *Gastroenterology* **2018**, *155*, 1716–1719.e4. [CrossRef] [PubMed]
7. Rea, D.; Coppola, G.; Palma, G.; Barbieri, A.; Luciano, A.; Del Prete, P.; Rossetti, S.; Berretta, M.; Facchini, G.; Perdonà, S.; et al. Microbiota effects on cancer: From risks to therapies. *Oncotarget* **2018**, *9*, 17915–17927. [CrossRef]
8. Barreto, S.G. We Asked the Experts: Providing the Road Map to Uncovering the Pathophysiology of Young-Onset Cancer to Guide Treatment and Preventive Strategies. *World J. Surg.* **2020**, *44*, 3212–3213. [CrossRef] [PubMed]
9. Arık, A.; Dodd, E.; Cairns, A.; Streftaris, G. Socioeconomic disparities in cancer incidence and mortality in England and the impact of age-at-diagnosis on cancer mortality. *PLoS ONE* **2021**, *16*, e0253854. [CrossRef]
10. Hoebel, J.; Kroll, L.E.; Fiebig, J.; Lampert, T.; Katalinic, A.; Barnes, B.; Kraywinkel, K. Socioeconomic Inequalities in Total and Site-Specific Cancer Incidence in Germany: A Population-Based Registry Study. *Front. Oncol.* **2018**, *8*, 402. [CrossRef] [PubMed]
11. Australian Institute of Health and Welfare. *The Health and Welfare of Australia's Aboriginal and Torres Strait Islander People, an Overview 2011*; Contract No Cat. No Ihw 42; AIHW: Canberra, Australia, 2011.
12. Ryder, C.; Mackean, T.; Hunter, K.; Towers, K.; Rogers, K.; Holland, A.J.A.; Ivers, R. Factors contributing to longer length of stay in Aboriginal and Torres Strait Islander children hospitalised for burn injury. *Inj. Epidemiol.* **2020**, *7*, 52. [CrossRef] [PubMed]
13. Devi, S. Native American health left out in the cold. *Lancet* **2011**, *377*, 1481–1482. [CrossRef]
14. Australian Institute of Health and Welfare. *Mortality and Life Expectancy of Indigenous Australian's 2008 to 2012*; Contract No Cat. No Ihw 140; AIHW: Canberra, Australia, 2014.
15. Hill, K.; Barker, B.; Vos, T. Excess Indigenous Mortality: Are Indigenous Australians More Severely Disadvantaged Than Other Indigenous Populations? *Int. J. Epidemiol.* **2007**, *36*, 580–589. [CrossRef]
16. Chong, A.; Roder, D. Exploring differences in survival from cancer among Indigenous and non-Indigenous Australians: Implications for health service delivery and research. *Asian Pac. J. Cancer Prev.* **2010**, *11*, 953–961. [PubMed]
17. Condon, J.R.; Barnes, T.; Armstrong, B.K.; Selva-Nayagam, S.; Elwood, J.M. Stage at diagnosis and cancer survival for Indigenous Australians in the Northern Territory. *Med. J. Aust.* **2005**, *182*, 277–280. [CrossRef] [PubMed]
18. Moore, S.P.; O'Rourke, P.K.; Mallitt, K.-A.; Garvey, G.; Green, A.C.; Coory, M.D.; Valery, P.C. Cancer incidence and mortality in Indigenous Australians in Queensland, 1997-2006. *Med. J. Aust.* **2010**, *193*, 590–593. [CrossRef] [PubMed]
19. Valery, P.C.; Coory, M.; Stirling, J.; Green, A. Cancer diagnosis, treatment, and survival in Indigenous and non-Indigenous Australians: A matched cohort study. *Lancet* **2006**, *367*, 1842–1848. [CrossRef]
20. Zhang, X.; Condon, J.R.; Rumbold, A.R.; Cunningham, J.; Roder, D.M. Estimating cancer incidence in Indigenous Australians. *Aust. N. Z. J. Public Health* **2011**, *35*, 477–485. [CrossRef] [PubMed]
21. Brands, J.; Garvey, G.; Anderson, K.; Cunningham, J.; Chynoweth, J.; Wallington, I.; Morris, B.; Knott, V.; Webster, S.; Kinsella, L.; et al. Development of a National Aboriginal and Torres Strait Islander Cancer Framework: A Shared Process to Guide Effective Policy and Practice. *Int. J. Environ. Res. Public Health* **2018**, *15*, 942. [CrossRef]

22. Condon, J.R.; Zhang, X.; Baade, P.; Griffiths, K.; Cunningham, J.; Roder, D.M.; Coory, M.; Jelfs, P.L.; Threlfall, T. Cancer survival for Aboriginal and Torres Strait Islander Australians: A national study of survival rates and excess mortality. *Popul. Health Metrics* **2014**, *12*, 1. [CrossRef]
23. Australian Institute of Health and Welfare. *Australia's Health 2016*; Australia's Health Series No. 15. Cat. No. AUS 199; AIHW: Canberra, Australia, 2016.
24. Australian Institute of Health and Welfare. *Profile of Indigenous Australians*; AIHW: Canberra, Australia, 2020.
25. Northern Territory Legislation. Cancer (Registration) Act. Available online: https://legislation.nt.gov.au/legislation/cancer-registration-act-2009 (accessed on 5 June 2022).
26. Breau, G.; Ellis, U. Risk Factors Associated with Young-Onset Colorectal Adenomas and Cancer: A Systematic Review and Meta-Analysis of Observational Research. *Cancer Control* **2020**, *27*, 1073274820976670. [CrossRef] [PubMed]
27. Kim, N.H.; Jung, Y.S.; Yang, H.-J.; Park, S.-K.; Park, J.H.; Park, D.I.; Sohn, C.I. Prevalence of and Risk Factors for Colorectal Neoplasia in Asymptomatic Young Adults (20–39 Years Old). *Clin. Gastroenterol. Hepatol.* **2018**, *17*, 115–122. [CrossRef] [PubMed]
28. Liu, P.-H.; Wu, K.; Ng, K.; Zauber, A.G.; Nguyen, L.; Song, M.; He, X.; Fuchs, C.S.; Ogino, S.; Willett, W.C.; et al. Association of Obesity with Risk of Early-Onset Colorectal Cancer Among Women. *JAMA Oncol.* **2019**, *5*, 37–44. [CrossRef] [PubMed]
29. Laine Tay, E.E.; Guthridge, S.; Zhang, X. Preventable Cancers in the Northern Territory—A Snapshot. *Chronicle* **2010**, *17*, 2–3.
30. Australian Bureau of Statistics. Overweight and Obesity, Australia. Available online: https://www.abs.gov.au/statistics/health/health-conditions-and-risks/overweight-and-obesity/latest-release (accessed on 5 June 2022).
31. Chakrabarti, S.; Peterson, C.Y.; Sriram, D.; Mahipal, A. Early Stage Colon Cancer: Current Treatment Standards, Evolving Paradigms, and Future Directions. *World J. Gastrointest. Oncol.* **2020**, *12*, 808–832. [CrossRef] [PubMed]
32. Lee, J.; Kim, I.-H.; Kim, J.S.; Kim, S.W.; Kim, J.G.; Oh, S.T.; Kang, W.K.; Lee, M.A. Different clinical characteristics in sporadic young-age onset colorectal cancer. *Medicine* **2016**, *95*, e4840. [CrossRef] [PubMed]
33. Bergquist, J.; Leiting, J.L.; Habermann, E.B.; Cleary, S.; Kendrick, M.L.; Smoot, R.L.; Nagorney, D.M.; Truty, M.J.; Grotz, T.E. Early-onset gastric cancer is a distinct disease with worrisome trends and oncogenic features. *Surgery* **2019**, *166*, 547–555. [CrossRef] [PubMed]
34. Piciucchi, M.; Capurso, G.; Valente, R.; Larghi, A.; Archibugi, L.; Signoretti, M.; Stigliano, S.; Zerboni, G.; Barucca, V.; La Torre, M.; et al. Early Onset Pancreatic Cancer: Risk Factors, Presentation and Outcome. *Pancreatology* **2015**, *15*, 151–155. [CrossRef] [PubMed]
35. Ntala, C.; Debernardi, S.; Feakins, R.M.; Crnogorac-Jurcevic, T. Demographic, clinical, and pathological features of early onset pancreatic cancer patients. *BMC Gastroenterol.* **2018**, *18*, 139. [CrossRef] [PubMed]
36. Lew, J.-B.; Feletto, E.; Worthington, J.; Roder, D.; Canuto, K.; Miller, C.; D'Onise, K.; Canfell, K. The potential for tailored screening to reduce bowel cancer mortality for Aboriginal and Torres Strait Islander peoples in Australia: Modelling study. *J. Cancer Policy* **2022**, *32*, 100325. [CrossRef] [PubMed]
37. Paulson, E.C.; Wirtalla, C.; Armstrong, K.; Mahmoud, N.N. Gender Influences Treatment and Survival in Colorectal Cancer Surgery. *Dis. Colon Rectum* **2009**, *52*, 1982–1991. [CrossRef] [PubMed]
38. Majek, O.; Gondos, A.; Jansen, L.; Emrich, K.; Holleczek, B.; Katalinic, A.; Nennecke, A.; Eberle, A.; Brenner, H.; Gekid Cancer Survival Working Group. Sex Differences in Colorectal Cancer Survival: Population-Based Analysis of 164,996 Colorectal Cancer Patients in Germany. *PLoS ONE* **2013**, *8*, e68077. [CrossRef] [PubMed]
39. Australian Bureau of Statistics. Smoking. Available online: https://www.abs.gov.au/statistics/health/health-conditions-and-risks/smoking/latest-release#data-download (accessed on 5 June 2022).
40. Alcohol. Available online: https://www.abs.gov.au/statistics/health/health-conditions-and-risks/alcohol-consumption/latest-release (accessed on 5 June 2022).
41. National Health Survey: First Results: Overweight and Obesity. 4364.0.55.001. 2019. Available online: https://iepcp.org.au/wp-content/uploads/2019/01/4364.0.55.001-national-health-survey-first-results-2017-18.pdf (accessed on 5 June 2022).
42. Ward, M.C.; Agarwal, A.; Bish, M.; James, R.; Faulks, F.; Pitson, J.; Yuen, N.; Mnatzaganian, G. Trends in obesity and impact on obstetric outcomes in a regional hospital in Victoria, Australia. *Aust. N. Z. J. Obstet. Gynaecol.* **2019**, *60*, 204–211. [CrossRef]
43. McCall, S.J.; Li, Z.; Kurinczuk, J.J.; Sullivan, E.; Knight, M. Binational Cohort Study Comparing the Management and Outcomes of Pregnant Women with a Bmi > 50–59.9 Kg/M(2) and Those with a Bmi ≥ 60 Kg/M(2). *BMJ Open* **2018**, *8*, e021055. [CrossRef] [PubMed]
44. Chu, Q.D.; Hsieh, M.-C.; Gibbs, J.F.; Wu, X.-C. Social determinants of health associated with poor outcome for rural patients following resected pancreatic cancer. *J. Gastrointest. Oncol.* **2021**, *12*, 2567–2578. [CrossRef] [PubMed]
45. Dhahri, A.; Kaplan, J.; Naqvi, S.M.H.; Brownstein, N.C.; Ntiri, S.O.; Imanirad, I.; Felder, S.I.; Dineen, S.P.; Sanchez, J.; Dessureault, S.; et al. The impact of socioeconomic status on survival in stage III colon cancer patients: A retrospective cohort study using the SEER census-tract dataset. *Cancer Med.* **2021**, *10*, 5643–5652. [CrossRef] [PubMed]
46. Anderson, I.; Crengle, S.; Kamaka, M.L.; Chen, T.H.; Palafox, N.; Jackson-Pulver, L. Indigenous Health in Australia, New Zealand, and the Pacific. *Lancet* **2006**, *367*, 1775–1785. [CrossRef]
47. Australian Government. Australian Instititute of Health and Welfare. Indigenous Life Expectancy and Deaths. Available online: https://www.aihw.gov.au/reports/australias-health/indigenous-life-expectancy-and-deaths (accessed on 9 April 2022).

48. Moore, S.P.; Antoni, S.; Colquhoun, A.; Healy, B.; Ellison-Loschmann, L.; Potter, J.D.; Garvey, G.; Bray, F. Cancer Incidence in Indigenous People in Australia, New Zealand, Canada, and the USA: A Comparative Population-Based Study. *Lancet Oncol.* **2015**, *16*, 1483–1492. [CrossRef]
49. Banham, D.; Roder, D.; Eckert, M.; Howard, N.J.; Canuto, K.; Brown, A. Cancer treatment and the risk of cancer death among Aboriginal and non-Aboriginal South Australians: Analysis of a matched cohort study. *BMC Health Serv. Res.* **2019**, *19*, 771. [CrossRef]
50. Diaz, A.; Whop, L.J.; Valery, P.C.; Moore, S.P.; Cunningham, J.; Garvey, G.; Condon, J.R. Cancer outcomes for Aboriginal and Torres Strait Islander Australians in rural and remote areas. *Aust. J. Rural Health* **2015**, *23*, 4–18. [CrossRef]
51. Cunningham, J.; Rumbold, A.R.; Zhang, X.; Condon, J. Incidence, aetiology, and outcomes of cancer in Indigenous peoples in Australia. *Lancet Oncol.* **2008**, *9*, 585–595. [CrossRef]
52. Siegel, J.; Engelhardt, K.E.; Hornor, M.A.; Morgan, K.A.; Lancaster, W.P. Travel distance and its interaction with patient and hospital factors in pancreas cancer care. *Am. J. Surg.* **2020**, *221*, 819–825. [CrossRef] [PubMed]
53. Moore, S.P.; Green, A.C.; Bray, F.; Garvey, G.; Coory, M.; Martin, J.; Valery, P.C. Survival disparities in Australia: An analysis of patterns of care and comorbidities among indigenous and non-indigenous cancer patients. *BMC Cancer* **2014**, *14*, 517. [CrossRef] [PubMed]
54. Walter, M. The Voice of Indigenous Data: Beyond the Markers of Disadvantage. *Griffith Rev.* **2018**, *60*, 256.
55. Walter, M.; Suina, M. Indigenous data, indigenous methodologies and indigenous data sovereignty. *Int. J. Soc. Res. Methodol.* **2018**, *22*, 233–243. [CrossRef]
56. Salmon, M.; Skelton, F.; Thurber, K.A.; Bennetts Kneebone, L.; Gosling, J.; Lovett, R.; Walter, M. Intergenerational and Early Life Influences on the Well-Being of Australian Aboriginal and Torres Strait Islander Children: Overview and Selected Findings from Footprints in Time, the Longitudinal Study of Indigenous Children. *J. Dev. Orig. Health Dis.* **2019**, *10*, 17–23. [CrossRef]

cancers

MDPI

Article

Intrinsic Cellular Susceptibility to Barrett's Esophagus in Adults Born with Esophageal Atresia

Chantal A. ten Kate [1,2,3], Annelies de Klein [3], Bianca M. de Graaf [3], Michail Doukas [4], Antti Koivusalo [5], Mikko P. Pakarinen [5], Robert van der Helm [3], Tom Brands [3], Hanneke IJsselstijn [1], Yolande van Bever [3], René M.H. Wijnen [1], Manon C.W. Spaander [2] and Erwin Brosens [3,*]

1 Department of Pediatric Surgery and Intensive Care Children, Erasmus MC-Sophia Children's Hospital, 3000 CA Rotterdam, The Netherlands; c.tenkate@erasmusmc.nl (C.A.t.K.); h.ijsselstijn@erasmusmc.nl (H.I.); r.wijnen@erasmusmc.nl (R.M.H.W.)
2 Department of Gastroenterology and Hepatology, Erasmus MC Cancer Institute, 3000 CA Rotterdam, The Netherlands; v.spaander@erasmusmc.nl
3 Department of Clinical Genetics, Erasmus MC Sophia Children's Hospital, 3000 CA Rotterdam, The Netherlands; a.deklein@erasmusmc.nl (A.d.K.); b.degraaf@erasmusmc.nl (B.M.d.G.); r.vanderhelm@erasmusmc.nl (R.v.d.H.); t.brands@erasmusmc.nl (T.B.); y.vanbever@erasmusmc.nl (Y.v.B.)
4 Department of Pathology, Erasmus MC, 3000 CA Rotterdam, The Netherlands; m.doukas@erasmusmc.nl
5 Department of Pediatric Surgery, University of Helsinki, Children's Hospital, 281, 000290 Helsinki, Finland; Antti.Koivusalo@hus.fi (A.K.); Mikko.Pakarinen@hus.fi (M.P.P.)
* Correspondence: e.brosens@erasmusmc.nl; Tel.: +31-10-70-37643

Simple Summary: We investigated the increased prevalence of Barrett's esophagus in adults with esophageal atresia. A higher polygenic risk score and disturbances in inflammatory, stress response and oncological pathways upon acid exposure suggest a genetic susceptibility and increased induction of inflammatory processes. Although further research is required to explore this hypothesis, this could be a first-step into selecting patients that are more at risk to develop Barrett's esophagus and/or esophageal carcinoma. Currently, an endoscopic screening and surveillance program is in practice in our institution for patients born with esophageal atresia, to early detect (pre)malignant lesions. Since recurrent endoscopies can be a burden for the patient, selecting patients by for example genetic susceptibility would allow to only include those at risk in future practice.

Abstract: The prevalence of Barrett's esophagus (BE) in adults born with esophageal atresia (EA) is four times higher than in the general population and presents at a younger age (34 vs. 60 years). This is (partly) a consequence of chronic gastroesophageal reflux. Given the overlap between genes and pathways involved in foregut and BE development, we hypothesized that EA patients have an intrinsic predisposition to develop BE. Transcriptomes of Esophageal biopsies of EA patients with BE (n = 19, EA/BE); EA patients without BE (n = 44, EA-only) and BE patients without EA (n = 10, BE-only) were compared by RNA expression profiling. Subsequently, we simulated a reflux episode by exposing fibroblasts of 3 EA patients and 3 controls to acidic conditions. Transcriptome responses were compared to the differential expressed transcripts in the biopsies. Predisposing single nucleotide polymorphisms, associated with BE, were slightly increased in EA/BE versus BE-only patients. RNA expression profiling and pathway enrichment analysis revealed differences in retinoic acid metabolism and downstream signaling pathways and inflammatory, stress response and oncological processes. There was a similar effect on retinoic acid signaling and immune response in EA patients upon acid exposure. These results indicate that epithelial tissue homeostasis in EA patients is more prone to acidic disturbances.

Keywords: acid sensitivity; genetic predisposition; esophageal carcinoma; inflammatory response; esophagitis

Citation: ten Kate, C.A.; de Klein, A.; de Graaf, B.M.; Doukas, M.; Koivusalo, A.; Pakarinen, M.P.; van der Helm, R.; Brands, T.; IJsselstijn, H.; van Bever, Y.; et al. Intrinsic Cellular Susceptibility to Barrett's Esophagus in Adults Born with Esophageal Atresia. *Cancers* **2022**, *14*, 513. https://doi.org/10.3390/cancers14030513

Academic Editor: Hidekazu Suzuki

Received: 4 December 2021
Accepted: 10 January 2022
Published: 20 January 2022

1. Introduction

Esophageal atresia (EA) is a congenital foregut malformation, of which improved survival rates have resulted in a growing adult population [1]. This raises new challenges in patient care as more emphasis is placed on long-term morbidities than short-term mortality. Respiratory and gastrointestinal symptoms require long-term follow-up [2]. Many adults born with EA (EA adults) suffer from chronic gastroesophageal reflux (GER), which is often underreported by patients due to an altered perception of discomfort [3]. GER can lead to reflux esophagitis, a nonspecific inflammation of the esophagus. Furthermore, the mucosal damage resulting from GER induces the replacement of esophageal squamous epithelium by gastric columnar epithelium containing goblet cells. This precursor lesion, intestinal metaplasia (IM) also known as Barrett's esophagus (BE), can develop via dysplasia into esophageal adenocarcinoma (EAC) [4]. Basal cells at the squamous-columnar junction are the origin of the BE cell population [5]. BE tissue has crypts composed of various combinations of goblet cells, mucinous cells, endocrine cells, enterocytes and Paneth cells [6]. The prevalence of BE in EA adults is 4–5 times higher than in the general population (6.6% vs. 1.6%), and presents at a much younger median age (34 vs. 60 years) [3]. In the Erasmus MC-Sophia Children's hospital cohort, EAC has been reported in three EA patients, and—surprisingly—also esophageal squamous cell carcinoma (ESCC) is seen more frequently in patients with EA at a younger age compared with the general population [3].

Disturbances in developmental signaling pathways are often associated with metaplasia and cancer transformation. The overlap of these pathways, disease genes and risk loci for foregut morphogenesis and BE development are suggestive of a shared etiology. During embryonic development the foregut separates into the future trachea and esophagus under the influence of spatiotemporal regulated transcriptional programs. These are regulated by gradients of morphogens that lay the blueprint for their interacting cells to develop into the various esophageal cell types and structures. Six intertwined pathways are crucial in this process: TGFB-BMP, Notch, FGF, WNT, Hedgehog and retinoic acid (RA) signaling [7]. TGFB-BMP signaling [8], SHH signaling [9] as well as RA signaling [10] are dysregulated in BE. Additionally, genome-wide association studies (GWAS) describe risk loci for the development of BE, EAC and ESCC near genes involved in these foregut developmental genes and pathways. These include *TBX5*, *GDF7*, *CRTC1*, *BARX1*, *FOXP1* and *FOXF1* [11].

Given the increased incidence of BE in EA adults, endoscopic surveillance is recommended [12]. Surveillance leads to early detection of BE or esophageal carcinoma, but could also create an unnecessary burden of repeated endoscopies for those not at risk as well as substantial added health care costs. Identifying patients at risk for developing BE could be a first step towards a tailor-made surveillance strategy. In this study, we hypothesize that patients born with EA have an increased (genetic) susceptibility for BE development. We aim to identify this predisposition by comparing risk loci burden and transcriptomes of patients with EA who have developed BE with EA patients without BE, and patients with BE without an EA history. We show that in both groups BE is histopathologically similar. However, the effect of acid reflux seems different with intrinsic cellular differences in inflammatory and stress response pathways, RA metabolism and signaling.

2. Materials and Methods

2.1. Study Population

Our institutional review board approved this case-control study (MEC-2018-1500). In our surveillance program, patients undergo upper endoscopies with histologic evaluation of biopsies taken according to a standardized protocol [3]. Biopsies and blood used in this study were retrieved from the Biobank Esophageal Atresia (MEC-2015-645) and the Biobank Barrett (MEC-2010-094). Mucosal esophageal biopsies were taken from two sites: (1) unaffected esophageal squamous cell epithelium (SQ), in EA patients taken above the original anastomosis; and (2) the GEJ or—if present—from Barrett's mucosa. Sample extraction protocol and storage are described in Supplementary Methods SM1. Additionally, we genotyped six EA/BE patients from a Finnish cohort study (447/E7/2005) [13], as well as

730 ancestry matched (broadly European) unaffected controls. For the in vitro experiments we used human fibroblasts from EA patients and healthy controls. EA fibroblast lines were taken during routine diagnostic procedures. Control fibroblast lines are anonymized lines that taken previously during unrelated routine diagnostic procedures and stored for research purposes. We compared three groups of patients: patients with EA who have developed BE (EA/BE), patients with EA without BE (EA-only), and patients with BE without EA in history (BE-only) BE-only patients were matched for age and gender with EA/BE patients. See Figure 1 for study set-up.

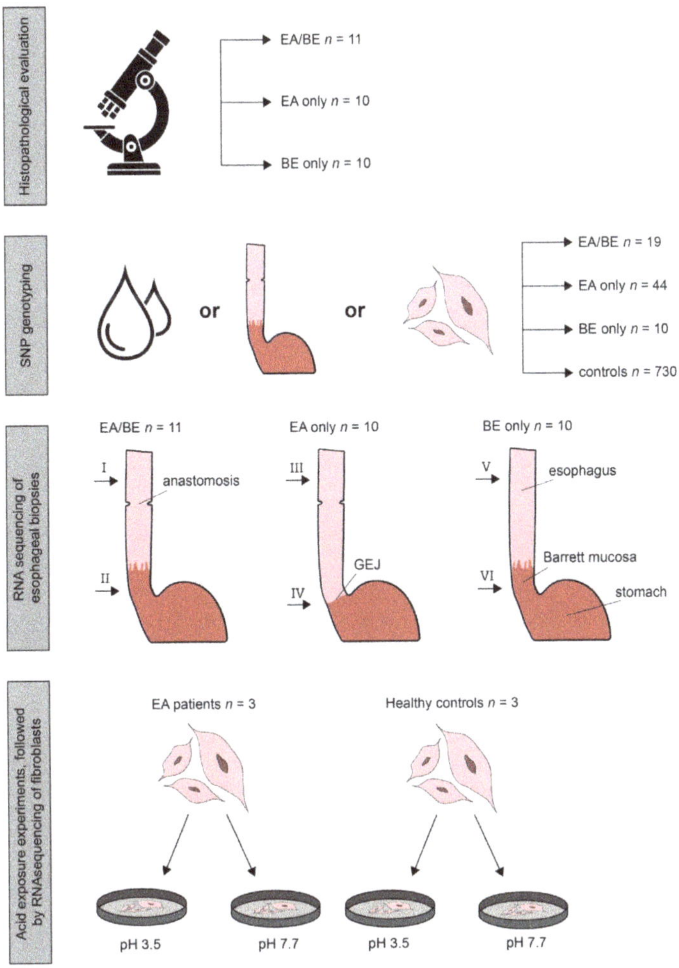

Figure 1. Schematic overview of the study set-up and number of patients included in each part. We compared three groups of patients: patients with esophageal atresia (EA) who have developed Barrett's esophagus (BE, EA/BE), patients with EA without BE (EA-only), and patients with BE without EA in history (BE-only). BE-only patients were matched for age and gender with EA/BE patients. Roman numerals I to VI indicate the subgroups, based on the location of the biopsies. GEJ = gastroesophageal junction.

2.2. Histopathological Evaluation

Hematoxylin and eosin-stained histological slides were retrieved from the archives of all patients of whom biopsies had been collected for RNA sequencing. All slides were blinded reassessed by a BE expert pathologist, according to a review-based checklist [6]. Potential differences were scored between the three groups.

2.3. SNP Genotyping and Calculation of Predisposing SNPs, Associated with BE

DNA extraction and quantification was done according standard procedures (see Supplementary Methods SM2). Processing of the SNP array genotyping chips (Infinium Global Screening Array v1.0 or v3.0 Illumina, Inc., San Diego, CA, USA) was done according to the manufacturer's standard protocol (SM3). Output was generated using Illumina Genome studio v2.0 (Illumina, San Diego, CA, USA). Predisposition loci (and corresponding lead or proxy SNPs) associated with BE, EAC and/or ESCC were derived from literature. We used genotype data from EA/BE patients ($n = 19$), EA-only patients ($n = 44$), BE-only patients ($n = 10$) and controls ($n = 730$) to see if previously BE associated SNPs were more prevalent in EA/BE patients (see Supplementary Methods SM3). We used the allele counts and published ORs of the associated SNPs to calculate a polygenic risk score (PGRS) using an additive model: $\text{PGRS} = \sum \text{Ln (OR risk allele)} \times \text{ allele count}$ (see Supplementary Tables S1–S5) Since we do not know if these ORs are precise enough to calculate the risk for the combination of EA and BE, we used the ORs of the associated SNPs calculated from our study population in a second calculation (see Supplementary Table S6). Using a Kruskal–Wallis test and Mann–Whitney tests, we compared the PGRS between the different groups. All statistical analyses were performed in SPSS V.25.0 (IBM, Chicago, IL, USA), with a significance level of $p < 0.05$.

2.4. RNA Sequencing, Differential Gene Expression and Pathway Enrichment Analysis

RNA extraction and quantification was done according standard procedures (see Supplementary Methods SM2). Genome-wide individual gene expression raw counts are available in Supplementary Datafile S1. Differential expression was calculated between (sub)groups (see Supplementary Methods SM4). Genes with a maximum group mean > 2, a fold change ≥ 1.5 and a false discovery rate (FDR) p-value < 0.05 were considered significantly differentially expressed. All differentially expressed genes per subgroup analysis were uploaded into the Ingenuity Pathway Analysis (IPA) software (Qiagen, Venlo, The Netherlands). Core analysis was performed for each (sub)group. A p-value of <0.05 and a Z-score of ≥ 2 were considered significant. Our ethics committee does not allow sharing of individual patient or control genotype information in the public domain, including sequencing reads.

2.5. Acid Exposure Experiments

In absence of available epithelial cells for in-vitro studies we used fibroblast. Activated fibroblasts generate extracellular matrix components and regulate inflammation [14]. There are several lines of evidence supporting a role for fibroblasts in BE proliferation and cancer [15,16]. To simulate a one-time acid reflux episode on RNA level, human fibroblasts from EA patients ($n = 3$) and healthy controls ($n = 3$) were exposed to pH adjusted cell culture medium conditions (see Supplementary Methods SM5). Hydrochloric acid was added to culture medium until the desired pH level was reached. Subsequently, cells were washed with phosphate buffered saline (PBS) and given standard medium. After 24 h, survival was measured (see Supplementary Table S7) with the TC20™ Automated Cell Counter (Bio-Rad Laboratories B.V., Veenendaal, The Netherlands). Cell morphology was evaluated (see Supplementary Figure S1) with the Olympus IX70-S8F Inverted Fluorescence Microscope (Olympus Corporation, Tokyo, Japan). RNA was isolated and sequenced as described in Supplementary Methods SM2 and SM4. Expression levels were compared with the RNA sequencing results of the esophageal biopsies.

2.6. Study Approval

The Medical Ethics Committee of the Erasmus Medical Center Rotterdam approved this study (MEC-2015-645, MEC-2010-094, MEC-2012-387). All authors had access to the study data and reviewed and approved the final manuscript.

3. Results

3.1. Study Population

Patient characteristics are depicted in Supplementary Tables S8 and S9. Histopathological assessment (see Supplementary Figure S2) of the biopsies is summarized in Supplementary Table S10. Columnar epithelium was present in all groups, except for two EA-only patients (see Supplementary Table S11). Since EA-only patients were selected as not having metaplasia in the distal esophagus at endoscopy, this means that most biopsies could contain part of the cardia as well. Neutrophil granulocytes were absent in the majority of EA-only patients, while a varying degree of nonspecific inflammatory cell infiltrate was present in most of them. Focusing on the characteristics of BE, IM with the presence of goblet cells was similarly present in EA/BE patients and BE-only patients. The amount of IM was larger in BE-only patients. No dysplasia was found in any of the samples.

3.2. SNP (Single Nucleotide Polymorphism) Genotyping

Given the limited sample size of our study population, we used ORs selected from literature to calculate the contribution of predisposing associated SNPs (polygenetic risk score, PGRS). Supplementary Table S1 depicts an overview of the included SNPs and ORs. Using these ORs, we found a median PGRS of 3.24 (range 1.39–4.68) for EA/BE patients, of 2.98 (1.19–4.74) for EA-only patients and of 2.63 (1.85–3.53) for BE-only patients. There were no statistical significant differences between these groups (Figure 2A, panel a, all $p > 0.05$). When using our own data, we did find significant differences in PGRS between these groups (Figure 2A, panel b). A higher risk allele frequency was found for EA/BE patients versus BE-only patients for rs3784262 near ALDH1A2 ($p = 0.017$), and a lower risk allele frequency of rs3072 near GDF7 ($p = 0.009$) (Figure 2B and Supplementary Table S3).

3.3. RNA Sequencing of Esophageal Biopsy Specimens

An average of 88,378,214 reads per sample were generated (62,471,354–165,874,334). Of these reads, 98% (94.9–98.4) aligned to the human reference genome. A total of 9752 transcripts had a mean expression of $\geq$2 RPKM and were considered expressed. See Supplementary Tables S12–S14 for the quality reports. PCA of the gene expression data confirmed clustering of the samples into the three groups (see Supplementary Figures S3 and S4). PCA and quality control procedures included the exclusion of two outliers (BBE-017 and BBE-079).

3.4. Differential Expression and Pathway Enrichment Analysis of Esophageal Biopsy Specimens

Seven known BE disease genes [11] were differentially expressed between EA-only patients and EA/BE or BE-only patients (Figure 3 and Supplementary Table S15). Enriched pathways between EA/BE patients and BE-only patients were involved in RA signaling, stress response and inflammatory pathways, and oncological processes (see Figure 4 and Supplementary Table S16).

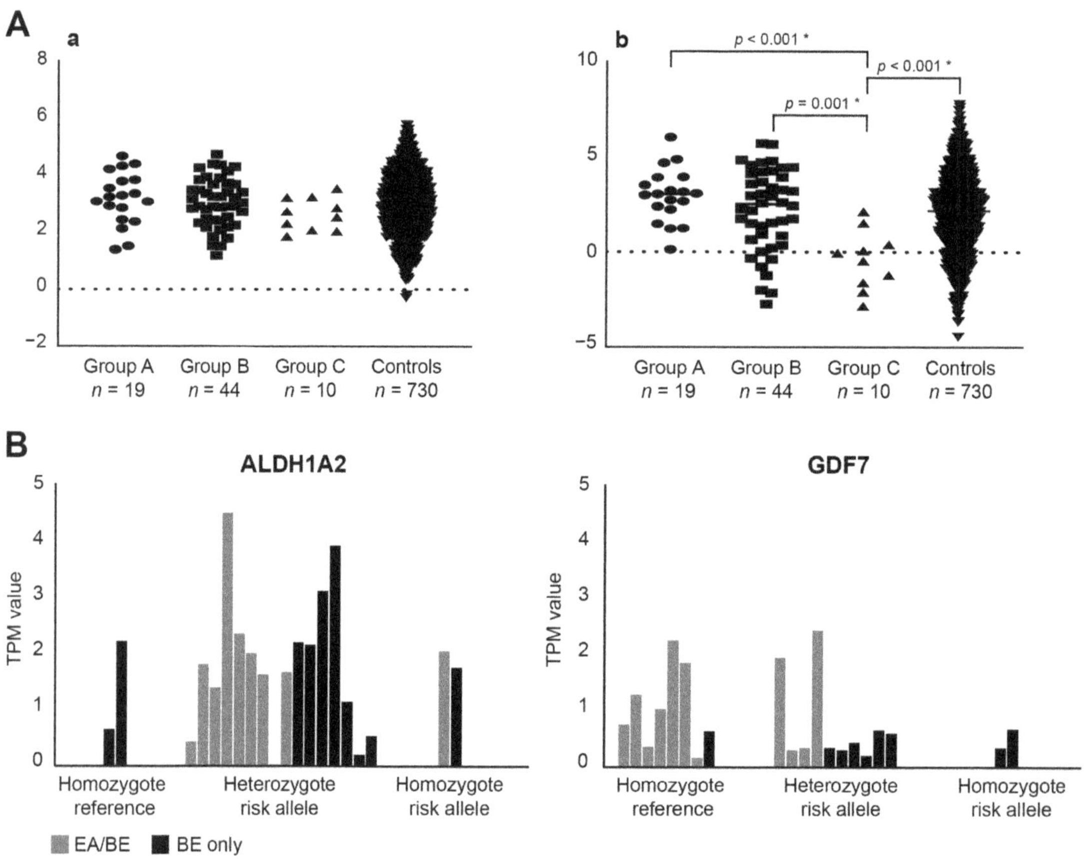

Figure 2. (**A**) Polygenic risk scores (PGRS) per patient. Group A = patients with esophageal atresia (EA) and Barrett's esophagus (BE), group B = patients with EA without BE, group C = patients with BE without EA in history. Panel a (left) are PGRS based on odds ratios (ORs) selected from the literature. No statistical significant differences between the groups were observed. Panel b (right): PGRS based on ORs calculated from our study population. We found a median PGRS of 3.05 (range 0.14–6.04) for EA/BE patients, of 2.52 (−2.73–5.72) for EA-only patients and of −0.24 (−2.83–2.15) for BE-only patients. A Kruskal–Wallis test revealed a significant difference in PGRS based on ORs calculated from our study population between the four groups ($p = 0.001$). T-statistics indicated a difference between BE-only patients versus EA/BE patients ($p < 0.001$), EA-only patients ($p = 0.001$) and controls ($p < 0.001$). Asterisk (*) indicates significance $p < 0.05$. (**B**) Gene expression levels for ALDH1A2 and GDF7 per patient, sorted based on the genotype of the patients. A higher risk allele frequency was found for EA/BE patients versus BE-only patients for rs3784262 near ALDH1A2 ($p = 0.017$) and a putative protective allele for rs3072 near GDF7 ($p = 0.009$). Looking at gene expression levels, GDF7 has slightly elevated TPM values for patients homozygote for the reference allele. No significant differences could be detected for these two associated SNPs. TPM = transcripts per million, EA = esophageal atresia, BE = Barrett's esophagus. Complete results can be found in Supplementary Tables S3–S6.

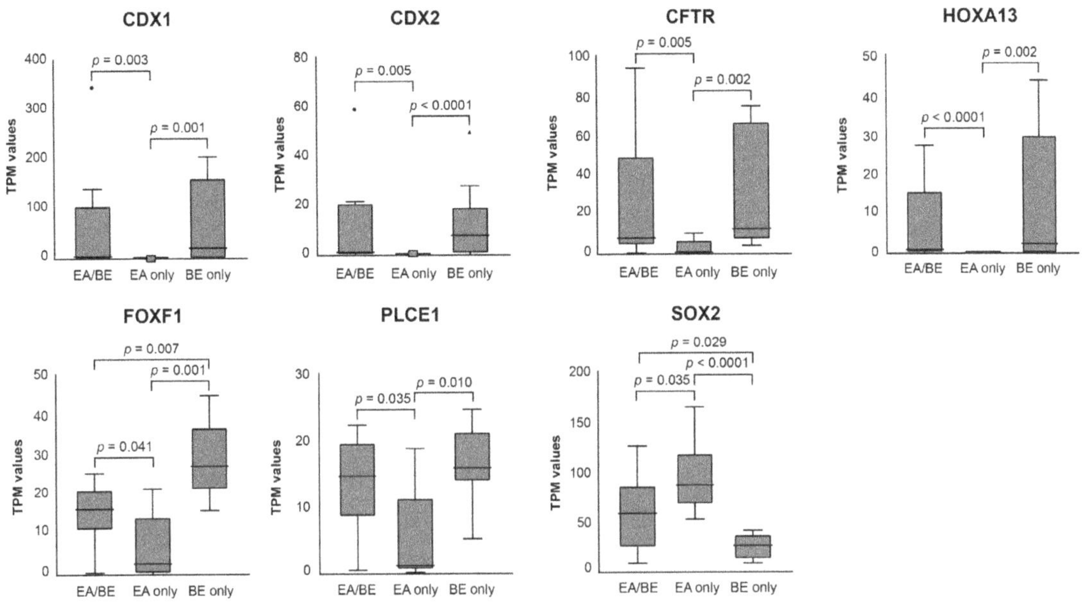

Figure 3. Gene expression levels per group for selected disease genes, involved in foregut morphogenesis and/or associated with Barrett's esophagus in literature, presented as median (interquartile range) with minimum and maximum values. We compared biopsies of the gastroesophageal junction between three groups of patients: patients with esophageal atresia (EA) who have developed Barrett's esophagus (BE) (EA/BE, $n = 11$), patients with EA without BE (EA-only, $n = 10$), and patients with BE without EA in history (BE-only, $n = 10$). TPM = transcripts per million, EA = esophageal atresia, BE = Barrett's esophagus.

3.5. Acid Exposure Experiments

To study the effect of GER on RNA level, we simulated a reflux episode in in vitro experiments (see Figure 1). First, we optimized the acid exposure experiment (see Supplementary Methods SM5). Next, we exposed fibroblasts from three EA patients and three healthy controls for 30 min to medium with pH 3.5 or to normal medium (control). Cells exposed to pH 3.5 showed cell rounding and irregular cell membranes (see Supplementary Figure S5). After acid exposure, there was a clear difference between upregulated and downregulated genes, both in patients and controls (see Supplementary Figure S6). Ten pathways were enriched with differentially expressed genes between patients and controls (see Supplementary Table S17), that contained 244 differentially expressed genes. Subtracting the genes that were also differentially expressed without acid exposure, 81 genes of interest remained (see Supplementary Figure S7). Pathway analysis of these 81 genes confirmed enrichment of pathways mostly involved in inflammatory processes (see Supplementary Table S18). Finally, we compared the results of the pathway analysis of the biopsies with those of the fibroblasts after acid exposure. Of the enriched pathways between GEJ samples of EA/BE patients and BE-only patients, 20 pathways were also enriched between fibroblasts of EA patients and controls after acid exposure (Table 1. In total, seven genes within these pathways were differentially expressed in both the GEJ samples and the acid-exposed fibroblasts (see Supplementary Figure S8).

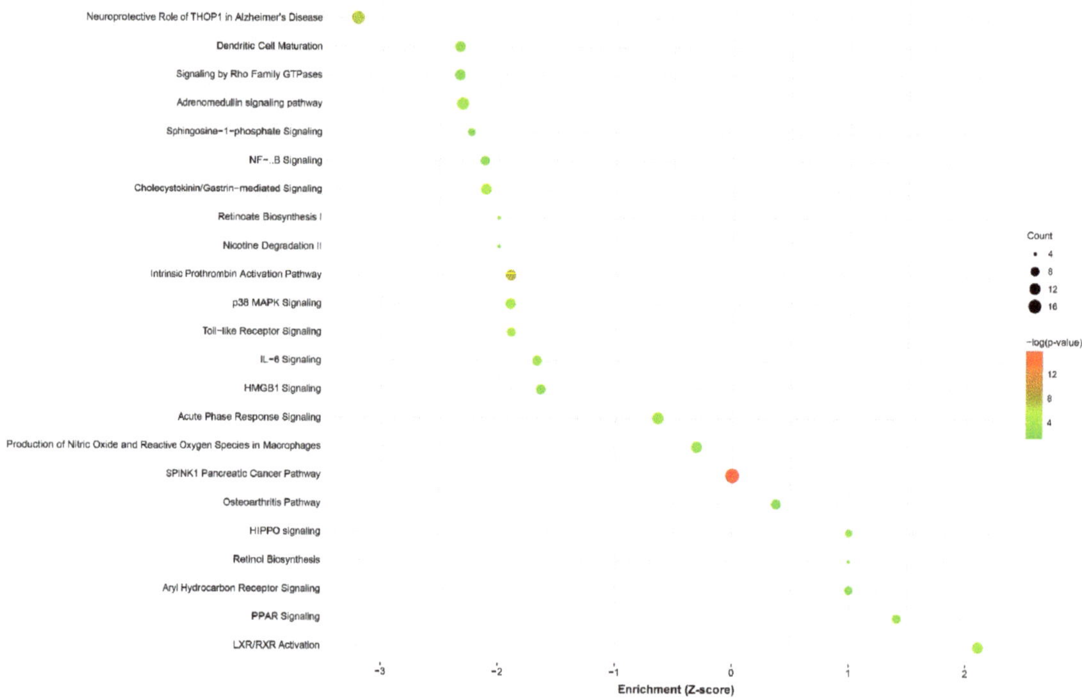

Figure 4. Bubble plot of canonical pathways, significantly enriched by differentially expressed genes, between gastroesophageal junction (GEJ) samples of group A (esophageal atresia (EA) with Barrett's esophagus (BE)) and GEJ samples of group C (BE-only). The color and size of the dots represent the range of the p-value and the number of molecultes mapped to the indicated pathways. Settings: p-value < 0.05 ($=-\log(p\text{-value}) > 1.3$), z-score < -2 or >2. SPINK1 Pancreatic Cancer Pathway is also the only significantly upregulated pathway, when comparing group A (EA/BE) with group C (BE-only). Plotted by http://www.bioinformatics.com.cn (accessed on 24 November 2021), a free online platform for data analysis and visualization.

Table 1. Overlap between canonical pathways, significantly enriched by differentially expressed genes, in the esophageal biopsy specimens versus the acid-exposed and non-exposed fibroblasts. II = gastroesophageal junction (GEJ) samples from group A (esophageal atresia (EA) and Barrett's esophagus (BE)), VI = GEJ samples from group C (BE-only). n = total number of canonical pathways significantly enriched by differentially expressed genes. N/A = not applicable, z-score could not be calculated. Grey box indicates that a pathway was not present in the results of that pathway analysis.

	Esophageal Biopsy Specimens		**Fibroblasts from Acid Exposure experiment**					
	II vs. VI (n = 353)		**EA Patients vs. Controls (Acid-Exposed) (n = 258)**		**EA Patients vs. Controls (Non-Exposed) (n = 314)**		**Acid-Exposed vs. Non-Exposed (All Samples) (n = 578)**	
Canonical Pathways	***−log(p-Value)***	***Z-Score***	***−log(p-Value)***	***Z-Score***	***−log(p-Value)***	***Z-Score***	***−log(p-Value)***	***Z-Score***
Agranulocyte Adhesion and Diapedesis	1.69	N/A	-	-	1.52	N/A	-	-
Altered T Cell and B Cell Signaling in Rheumatoid Arthritis	3.22	N/A	2.57	N/A	2.05	N/A	-	-
Atherosclerosis Signaling	4.93	N/A	2.04	N/A	2.23	N/A	-	-
Cholecystokinin/Gastrin-mediated Signaling	4.38	2.111	2.35	0	1.39	N/A	-	-
Communication between Innate and Adaptive Immune Cells	2.39	N/A	2.47	N/A	-	-	-	-
Dendritic Cell Maturation	2.27	2.333	4.600	−0.707	2.19	−1.633	-	-
Extrinsic Prothrombin Activation Pathway	1.36	N/A	2.34	N/A	-	-	-	-
Glucocorticoid Receptor Signaling	4.51	N/A	1.53	N/A	2.03	N/A	-	-
Graft-versus-Host Disease Signaling	4.23	N/A	3.600	N/A	-	-	-	-
HMGB1 Signaling	2.09	1.633	2.37	N/A	-	-	-	-
IL-6 Signaling	2.89	1.667	1.33	N/A	-	-	5.02	2.117
Intrinsic Prothrombin Activation Pathway	7.92	1.897	2.61	N/A	-	-	-	-
LXR/RXR Activation	4.31	−2.111	2.12	−1	-	-	-	-
MSP-RON Signaling Pathway	4.58	N/A	1.44	N/A	1.81	N/A	-	-
Osteoarthritis Pathway	1.44	−0.378	-	-	6.98	−1.265	-	-
PPAR Signaling	2.81	−1.414	2.33	0	-	-	2.57	−2.524
Production of Nitric Oxide and Reactive Oxygen Species in Macrophages	2.68	0.302	1.48	N/A	-	-	-	-
Retinol Biosynthesis	1.95	−1	1.49	N/A	-	-	-	-
Role of Macrophages, Fibroblasts and Endothelial Cells in Rheumatoid Arthritis	2.59	N/A	3.710	N/A	3.87	N/A	-	-
Role of Osteoblasts, Osteoclasts and Chondrocytes in Rheumatoid Arthritis	3.10	N/A	3.280	N/A	4.53	N/A	-	-
Sphingosine-1-phosphate Signaling	1.44	−2.236	1.40	N/A	2.38	−1	-	-

4. Discussion

In this first translational case-control study in adults born with esophageal atresia (EA), we compared EA patients who developed Barrett's esophagus (BE, EA/BE) to EA patients who did not develop BE (EA-only) and BE patients without a history of EA (BE-only). Previous studies described an increased prevalence of BE in EA adults—and at a much younger age—compared with the general population [3]. Over the years, several risk loci associated with BE and/or esophageal carcinoma have been published, of which many near genes involved in foregut development [11] (S2). This overlap made us hypothesize that EA patients have an increased (genetic) susceptibility to develop BE.

4.1. BE Characteristics of EA/BE Patients and BE-Only Patients

There is a twenty-year difference in the age at which biopsies were taken between EA/BE patients and BE-only patients. We confirmed the lack of morphological differences between these two groups. Although endoscopic esophagitis was absent in the majority of the BE-only patients, neutrophil granulocytes were present in these patients. The typical characteristics of BE (columnar metaplasia with presence of goblet cells) were equally present, although the larger amount of IM in BE-only patients is indicative of a more advanced stage. Paneth cells were present in some patients of both groups, a variety more often reported in BE [6].

4.2. The Contribution of BE Associated SNPs in EA/BE Patients

The overlap of genes involved in foregut development and risk loci for BE insinuates a genetic predisposition for EA patients to develop BE. For example, *FOXF1*, which is expressed in the developing foregut [17], *BARX1*, which is expressed at the tracheoesophageal separation site and inhibits Wnt signaling [18], and *FOXP1*, which regulates esophageal muscle development [19], have all been associated with BE in previous GWAS studies [11]. *FOXP1* has also been implicated as a tumor suppressor gene in several tissues including the gastrointestinal tract [20]. The ORs of these risk loci were often small and the GWAS studies included large sets of BE patients in order to detect these predispositions.

Regardless, there seems to be an elevated risk for EA patients. EA/BE patients have a higher median PGRS compared with BE-only patients (3.24 vs. 2.63, $p = 0.069$), which was confirmed and reached significance when using ORs calculated from our study population ($p < 0.001$, see Figure 2A and Supplementary Tables S4 and S6). Despite the small cohorts, the higher PGRS in EA/BE patients is suggestive for an increased predisposition, and a possible contribution for the earlier age of onset of BE in these patients. Such a relationship (higher PGRS and earlier disease onset) has been demonstrated previously in patients with atrial fibrillation [21]. However, differences in PGRS are not likely to be sufficient on their own to exclude EA patients from (pre)malignant screening protocols. Ideally, a screening algorithm would contain multiple risk factors of which the PGRS could be one. Further research would be required to confirm the impact of risk loci for BE and their potential benefit in surveillance strategies for EA patients.

Two predisposing associated SNPs proved enriched when comparing EA/BE patients with BE-only patients: rs3784262 near *ALDH1A2* (OR 3.94, $p = 0.017$) and rs3072 near *GDF7* (OR 0.22, $p = 0.009$). *ALDH1A2* (also known as *RALDH2*) is an enzyme that catalyzes the transformation of retinaldehyde into RA, a key morphogen in foregut development [22]. Lack of RA signaling results in increased TGFB-BMP signaling and hampers lung bud induction [23]. In contrast, BE is characterized by a higher expression of this enzyme, resulting in higher levels of RA [24]. *GDF7* is also a component of the TGFB-BMP signaling pathway. TGFB-BMP signaling is essential in esophageal formation by inhibiting *SOX2* in the ventral foregut [25] but also contributes to the differentiation of columnar epithelium and BE development by interacting with *CDX1* and *CDX2* [26]. Interestingly, the associated SNP *GDF7* seems a protective locus in EA/BE patients (OR 0.22, $p = 0.009$). The trends shown by these results are illustrative but more research is needed. Though EA/BE patients could have an increased genetic risk, the current sample sizes do not allow to draw firm conclusions.

4.3. EA/BE Patients Have Comparable Gene Expression of BE Disease Genes as BE-Only Patients

An earlier age of BE onset in EA patients could mean that epithelial homeostasis in these patients is more prone to disturbances. To investigate this, we sequenced RNA extracted from esophageal biopsies of three groups (EA/BE, EA-only and BE-only). We evaluated the expression of BE disease genes but found no difference in expression between EA/BE patients and BE-only patients. In both groups, these genes were upregulated compared to EA-only patients, indicating that the BE found in EA/BE patients is similar to the BE in BE-only patients.

4.4. EA/BE Patients Have an Increased Inflammatory Response

Since the expression of disease genes could not explain the earlier age of onset, we explored the complete transcriptome and corresponding differentially expressed genes and pathways. Many of the enriched pathways in EA/BE patients compared with BE-only patients, hinted at upregulated inflammatory (e.g., IL-6 signaling) and stress response pathways, downregulated oncological processes and dysregulated RA signaling (see Supplementary Table S16). Inflammatory cells produce carcinogenic compounds that can initiate DNA damage. The secretion of growth factors and cytokines increase proliferation and transition to tumor cells [27]. *SPINK1* expression itself has the potential to be a BE biomarker as it lacks expression in unaffected esophageal tissue (see Supplementary Figure S9).

Human studies and in vitro experiments have shown that exposure of esophageal tissue to low pH and/or bile acids may induce cell proliferation and reduce cell apoptosis through an increased expression of cyclo-oxygenase-2 (COX-2), prostaglandin E_2 (PGE2), mitogen-activated protein kinase (MAPK) and NF-κB pathways [28–31]. In our data, p38 MAPK Signaling and NF-κB Signaling are upregulated in EA/BE patients compared with BE-only patients. Given their proliferative and anti-apoptotic role, these pathways could be valuable for BE staging. Quante and coworkers showed that transgenic mice, overexpressing human *IL-1β*, presented with chronic inflammation, BE and esophageal dysplasia. Oral exposure to bile acids led to elevated IL-6 levels, accelerating BE development and progression into EAC, and implicating an IL-1β-IL-6 signalling cascade [32]. Clinical management of BE is focused around chemical inhibition of acid exposure and decrease of inflammation. Inhibition of gastric acid secretion with proton pump inhibitors (PPIs) reduces the transition to dysplasia in BE patients [33] and a combination of non-steroidal anti-inflammatory drugs (NSAIDs) and statins may reduce neoplastic progression [34]. Recently, it has been shown that the combination of high-doses esomeprazole and aspirin reduces high-grade dysplasia and EAC in BE patients [35]. Given the potentially altered response to acid in EA patients, the effectiveness of PPIs and NSAIDs in this population warrant further investigation.

Furthermore, stress response pathways are upregulated. Cholecystokinin/Gastrin-mediated Signaling is an activator of actin stress fiber formation and intertwined with stress response pathways as p38 MAPK Signaling, Sphingosine-1-phosphate Signaling and Signaling by Rho Family GTPases. These processes may lead to the conversion of squamous epithelium to columnar metaplasia. Another study showed that low pH and/or bile acids can induce oxidative stress, which causes DNA damage [36]. In combination with reduced apoptosis this can lead to dysplasia. When this is followed by neoplastic progression BE can develop into EAC.

4.5. Dysregulation of RA Metabolism and Signaling

RA is increased in BE and works—like bile acids—through the RXR receptors to transform squamous epithelium to columnar epithelium [10]. LXR/RXR activation, involved in RA mediated gene activation, is downregulated in EA/BE patients compared with BE-only patients. Retinol biosynthesis is also downregulated, whilst its downstream processes in all trans RA synthesis (Retinoate Biosynthesis I) are upregulated. Peroxisome proliferator-activated receptors (PPARs) are transcription factors activated by RA, generally upregulated in BE [37], but downregulated in EA/BE patients. Like discussed above, the downregulation of these pathways could indicate that BE-only patients are at a more advanced stage than EA/BE patients. Given the clinical differences (age and length of BE) between these patients, this does make sense.

4.6. Downregulation of the Hippo/YAP Pathway

Downregulation of oncological pathways in EA/BE patients could be indicative of either a decreased progression rate to dysplasia or a less advanced state of progression compared with BE-only patients. The Hippo/YAP pathway is important in cell proliferation, survival, and differentiation. Yes-association protein (YAP) expression is associated with dysplasia and adenocarcinoma [38]. Hippo signaling is involved in cell contact inhibition [39] as is Aryl Hydrocarbon Receptor Signaling [40]. Hippo activation (and YAP inactivation) is necessary for programmed cell death after detachment from the extracellular matrix [41]. Therefore, downregulation of this pathway could (in theory) decrease anoikis and increase the risk of tumor cell metastasis.

4.7. EA Patients Seem to Be More Sensitive to Acid Reflux Exposure

EA patients are earlier in life and more frequently exposed to GER. Chronic GER could be a consequence of the surgical repair: the lower esophageal sphincter is often retracted above the diaphragm, resulting in the loss of the natural reflux barrier function of the GEJ [42]. Other factors contributing to GER are impaired motility, delayed bolus clearance and delayed gastric emptying [43]. There seems to be a direct relationship of these symptoms with EA, as Adriamycin induced EA rats have impaired esophageal relaxation and a decreased number of ganglia and nerve fibers in the esophageal myenteric plexus [44]. The prevalence of mucosal damage is related to the level of pH exposure and to the composition of the acid reflux [45]. Animal studies have shown that acid fluids can activate pepsin, which inflicts injury and leads to mucosal damage [46].

We speculated that GER could result in an upregulation of inflammatory pathways. Additionally, EA patients could have a predisposition that makes them more sensitive to acid reflux than the general population. To explore these hypotheses, we performed in vitro experiments to simulate a one-time reflux episode in fibroblasts of EA patients and healthy controls. The enriched pathways of the GEJ biopsies of EA/BE patients showed an overlap with the enriched pathways of the fibroblasts of EA patients after acid exposure—but not with those of healthy controls. These overlapping pathways were again mostly involved inflammatory or oncological processes. For example, LXR/RXR Activation, PPAR Signaling and Retinol Biosynthesis were also enriched in fibroblasts of EA patients after acid exposure, hinting at intrinsic disturbances of RA signaling in EA patients under the influence of GER.

We do not know of the three patients used in the in vitro experiment will develop BE in time as the fibroblasts are derived of patients currently aged 29, 30 and 39 years old. It is, however, interesting that we could detect a similar predisposition in just 3 EA patients, and as a general response (in fibroblasts) to acid.

4.8. Strengths and Limitations

The main strength of this study is the broad investigative approach by combining histology, genotype, transcriptome and in vitro results. Some limitations should be addressed. First, due to the relative low incidence of EA and corresponding small sample sizes, we mostly observed trends and more EA/BE patients are needed to draw more robust conclusions. At this point, the difference in gene expression between EA/BE patients and EA-only patients is negligible. This could be due to the fact that most biopsies could contain part of the cardia. However, the power would increase substantially if we would know which EA patients have not developed BE throughout their life, as the current EA-only population is a mixture of patients who have not yet and will never develop BE. Second, EA is a heterogeneous disease. Our study population included both patients with isolated EA and patients with syndromes or multiple anomalies. This phenotypic heterogeneity might also be the results of a genetic heterogeneity. Thirdly, BE can present as a heterogeneous metaplastic mosaic, consisting of multiple individual crypts that arose from independent clones [47], which have distinct ploidies, copy number variations (CNV) and point mutations [48]. Heterogeneity in these crypts pose a risk of sampling error. Even within long segment BE, IM can be focally distributed [49]. Recent progress in genetic analysis of BE stem cells and EAC indicates that there are patient-specific driver genes affected in both the precursor lesion [50] and subsequent cancer of the esophagus [51]. Perhaps the heterogeneous background of de novo mutations [52] and de novo CNVs [53] in EA contributes to this patient-centred susceptibility. This could have created larger variances in gene expression per evaluated group. Subsequent experiments using single-cell sequencing of definite IM could reveal differences between patients that cannot be detected in whole biopsy specimens. Lastly, morphological differences were absent. However, segment length differences could be related to a difference in disease stage [54] and impact gene networks are prone to disturbances.

5. Conclusions

Altered regulation of p38 MAPK, NF-κB and RA signaling could have implications for (or be related to) the dysplastic progression. If Hippo/YAP signaling remains downregulated upon progression to cancer, the metastasis risk could be higher in EA patients due to reduced anoikis. An increased PGRS and upregulation of inflammatory pathways hint at a multifactorial contribution underlying the earlier age of onset of BE in EA patients. We did not evaluate mechanical factors such as loss of the natural reflux barrier due to the surgical repair and clinical factors such as impaired esophageal motility. These factors increase the level of acid exposure and likely add to the effect of risk loci and primed inflammatory pathways.

Supplementary Materials: The following are available online at https://www.mdpi.com/article/10.3390/cancers14030513/s1, Table S1: Genes and polymorphisms associated with Barrett's esophagus (BE), with selected proxy SNPs used for SNP array genotyping. Table S2: Genes and polymorphisms associated with Barrett's esophagus (BE), esophageal adenocarcinoma (EAC) or esophageal squamous cell carcinoma (ESCC). Table S3: Odds ratios calculated from the single nucleotide polymorphism (SNP) genotyping data. Table S4: Overview of polygenic risk scores (PGRS) for all groups. Table S5: Overview of the selected odds ratios (OR) used for the polygenic risk score. Table S6: Comparison of all groups separately for the polygenic risk score (PGRS). Table S7: Overview of survival rates of fibroblast cells after exposure to pH adjusted medium. Table S8: Basic characteristics of selected patients and controls for RNA sequencing of the esophageal biopsy specimen and for the SNP array genotyping. Table S9: Phenotype description. Table S10: Summarized results of reassessments of pathology slides of esophageal biopsy specimens. Table S11: Complete results of reassessments of pathology slides of esophageal biopsy specimens. Table S12: Results of RNA and DNA isolation. Table S13: Quality report of RNA sequencing data from esophageal biopsy specimens. Table S14: Results of RNA isolation from fibroblast of the acid exposure experiments. Table S15: Number of significantly differently expressed genes when comparing the different subgroups. Table S16: Canonical pathways, significantly enriched by differentially expressed genes, and corresponding diseases and bio functions. Table S17: Canonical pathways, enriched by differentially expressed genes. Table S18: Canonical pathways, significantly enriched by differentially expressed genes. Figure S1. Overview of survival rates of fibroblast cells after exposure to pH adjusted medium. Figure S2: Examples of the review-based checklist used for the histopathological assessments. Figure S3: Two-dimensional scatter plot of principal component analysis (PCA). Figure S4: Tree-dimensional scatter plot of principal component analysis (PCA). Figure S5: Morphology (20× magnification) of fibroblast cells. Figure S6: Heat map of mean transcript per million (TPM) for fibroblast cells after the in vitro experiment. Figure S7: Number of differentially expressed genes between patients with EA and healthy controls. Figure S8: Violin plots of gastroesophageal (GEJ) samples and box plots of acid-exposed fibroblasts. Figure S9. SPINK1 is a potential BE biomarker. Datafile S1: Individual gene counts (Excel sheet). SM1: Sample extraction protocol and storage. SM2: RNA and DNA isolation. SM3: SNP genotyping. SM4: RNA sequencing. SM5: Acid exposure experiments. References [55–78] are cited in the supplementary materials.

Author Contributions: Conceptualization, E.B., A.d.K. and M.C.W.S.; methodology, C.A.t.K. and E.B.; software, R.v.d.H., T.B.; validation, C.A.t.K. and B.M.d.G.; formal analysis, C.A.t.K. and E.B.; investigation, C.A.t.K. and E.B.; resources, M.D., A.K., M.P.P. and Y.v.B.; data curation, E.B.; writing—original draft preparation, C.A.t.K. and E.B.; writing—review and editing, C.A.t.K., A.d.K., R.v.d.H., M.C.W.S., Y.v.B., R.M.H.W., H.I.; visualization, C.A.t.K.; supervision, E.B.; project administration, E.B.; funding acquisition, M.C.W.S., R.M.H.W. and E.B. All authors have read and agreed to the published version of the manuscript.

Funding: This research received no external funding.

Institutional Review Board Statement: The study was conducted according to the guidelines of the Declaration of Helsinki, and approved by the Institutional Review Board of the Erasmus Medical Center Rotterdam (MEC-2015-645, MEC-2010-094, MEC-2012-387).

Informed Consent Statement: Written informed consent was obtained from all subjects involved in the study. Anonymized control lines were obtained from the cell repository of the department of Clinical Genetics, Erasmus MC, Rotterdam. The Erasmus MC Rotterdam has an opt-out procedure for the anonymous use of specified medical leftover material like blood, tissues, medical images and medical data.

Data Availability Statement: All transcriptome count data is in the supporting information files. Our ethics committee does not allow sharing of individual patient or control genotype information in the public domain.

Acknowledgments: In memoriam of our friend and colleague, we would like to dedicate this manuscript to Robert M.W. Hofstra. His insightful guidance during this project was always appreciated. This research project is supported by the European Reference Network for rare Inherited and Congenital Anomalies (ERNICA, https://ern-ernica.eu/, accessed on 24 November 2021). Furthermore, we thank all patients who participated in the Biobank Esophageal Atresia and Biobank Barrett for their cooperation. We thank Floor Vergouwe, Ruben van der Bogt and Anne-Fleur van Hal for their help collecting the patient material for the biobanks. We thank Hedwika Nastiti for her help with the sequencing work. We thank Jonathan Windster for his help with the graphic artwork.

Conflicts of Interest: The authors declare no conflict of interest.

References

1. Wang, B.; Tashiro, J.; Allan, B.J.; Sola, J.E.; Parikh, P.P.; Hogan, A.R.; Neville, H.L.; Perez, E.A. A nationwide analysis of clinical outcomes among newborns with esophageal atresia and tracheoesophageal fistulas in the United States. *J. Surg. Res.* **2014**, *190*, 604–612. [CrossRef] [PubMed]
2. Van Lennep, M.; Singendonk, M.M.J.; Dall'Oglio, L.; Gottrand, F.; Krishnan, U.; Terheggen-Lagro, S.W.J.; Omari, T.I.; Benninga, M.A.; van Wijk, M.P. Oesophageal atresia. *Nat. Rev. Dis. Primers* **2019**, *5*, 26. [CrossRef]
3. Vergouwe, F.W.T.; IJsselstijn, H.; Biermann, K.; Erler, N.S.; Wijnen, R.M.H.; Bruno, M.J.; Spaander, M.C.W. High Prevalence of Barrett's Esophagus and Esophageal Squamous Cell Carcinoma After Repair of Esophageal Atresia. *Clin. Gastroenterol. Hepatol.* **2018**, *16*, 513–521.e6. [CrossRef]
4. Solaymani-Dodaran, M.; Logan, R.F.; West, J.; Card, T.; Coupland, C. Risk of oesophageal cancer in Barrett's oesophagus and gastro-oesophageal reflux. *Gut* **2004**, *53*, 1070–1074. [CrossRef]
5. Jiang, M.; Li, H.; Zhang, Y.; Yang, Y.; Lu, R.; Liu, K.; Lin, S.; Lan, X.; Wang, H.; Wu, H.; et al. Transitional basal cells at the squamous-columnar junction generate Barrett's oesophagus. *Nature* **2017**, *550*, 529–533. [CrossRef] [PubMed]
6. Naini, B.V.; Souza, R.F.; Odze, R.D. Barrett's Esophagus: A Comprehensive and Contemporary Review for Pathologists. *Am. J. Surg. Pathol.* **2016**, *40*, e45–e66. [CrossRef]
7. Han, L.; Chaturvedi, P.; Kishimoto, K.; Koike, H.; Nasr, T.; Iwasawa, K.; Giesbrecht, K.; Witcher, P.C.; Eicher, A.; Haines, L.; et al. Single cell transcriptomics identifies a signaling network coordinating endoderm and mesoderm diversification during foregut organogenesis. *Nat. Commun.* **2020**, *11*, 4158. [CrossRef] [PubMed]
8. Milano, F.; van Baal, J.W.; Buttar, N.S.; Rygiel, A.M.; de Kort, F.; DeMars, C.J.; Rosmolen, W.D.; Bergman, J.J.; van Marle, J.; Wang, K.K.; et al. Bone morphogenetic protein 4 expressed in esophagitis induces a columnar phenotype in esophageal squamous cells. *Gastroenterology* **2007**, *132*, 2412–2421. [CrossRef]

9. Wang, D.H.; Clemons, N.J.; Miyashita, T.; Dupuy, A.J.; Zhang, W.; Szczepny, A.; Corcoran-Schwartz, I.M.; Wilburn, D.L.; Montgomery, E.A.; Wang, J.S.; et al. Aberrant epithelial-mesenchymal Hedgehog signaling characterizes Barrett's metaplasia. *Gastroenterology* **2010**, *138*, 1810–1822. [CrossRef]
10. Chang, C.L.; Lao-Sirieix, P.; Save, V.; De La Cueva Mendez, G.; Laskey, R.; Fitzgerald, R.C. Retinoic acid-induced glandular differentiation of the oesophagus. *Gut* **2007**, *56*, 906–917. [CrossRef]
11. Gharahkhani, P.; Fitzgerald, R.C.; Vaughan, T.L.; Palles, C.; Gockel, I.; Tomlinson, I.; Buas, M.F.; May, A.; Gerges, C.; Anders, M.; et al. Genome-wide association studies in oesophageal adenocarcinoma and Barrett's oesophagus: A large-scale meta-analysis. *Lancet Oncol.* **2016**, *17*, 1363–1373. [CrossRef]
12. Krishnan, U.; Mousa, H.; Dall'Oglio, L.; Homaira, N.; Rosen, R.; Faure, C.; Gottrand, F. Espghan-Naspghan Guidelines for the Evaluation and Treatment of Gastrointestinal and Nutritional Complications in Children with Esophageal Atresia-Tracheoesophageal Fistula. *J. Pediatric Gastroenterol. Nutr.* **2016**, *63*, 550–570. [CrossRef] [PubMed]
13. Koivusalo, A.I.; Pakarinen, M.P.; Lindahl, H.G.; Rintala, R.J. Endoscopic Surveillance after Repair of Oesophageal Atresia: Longitudinal Study in 209 Patients. *J. Pediatric Gastroenterol. Nutr.* **2016**, *62*, 562–566. [CrossRef] [PubMed]
14. Kato, T.; Noma, K.; Ohara, T.; Kashima, H.; Katsura, Y.; Sato, H.; Komoto, S.; Katsube, R.; Ninomiya, T.; Tazawa, H.; et al. Cancer-Associated Fibroblasts Affect Intratumoral CD8(+) and FoxP3(+) T Cells Via IL6 in the Tumor Microenvironment. *Clin. Cancer Res.* **2018**, *24*, 4820–4833. [CrossRef]
15. Wang, J.; Zhang, G.; Wang, J.; Wang, L.; Huang, X.; Cheng, Y. The role of cancer-associated fibroblasts in esophageal cancer. *J. Transl. Med.* **2016**, *14*, 30. [CrossRef]
16. Okawa, T.; Michaylira, C.Z.; Kalabis, J.; Stairs, D.B.; Nakagawa, H.; Andl, C.D.; Johnstone, C.N.; Klein-Szanto, A.J.; El-Deiry, W.S.; Cukierman, E.; et al. The functional interplay between EGFR overexpression, hTERT activation, and p53 mutation in esophageal epithelial cells with activation of stromal fibroblasts induces tumor development, invasion, and differentiation. *Genes Dev.* **2007**, *21*, 2788–2803. [CrossRef]
17. Mahlapuu, M.; Enerback, S.; Carlsson, P. Haploinsufficiency of the forkhead gene Foxf1, a target for sonic hedgehog signaling, causes lung and foregut malformations. *Development* **2001**, *128*, 2397–2406. [CrossRef]
18. Woo, J.; Miletich, I.; Kim, B.M.; Sharpe, P.T.; Shivdasani, R.A. Barx1-mediated inhibition of Wnt signaling in the mouse thoracic foregut controls tracheo-esophageal septation and epithelial differentiation. *PLoS ONE* **2011**, *6*, e22493. [CrossRef] [PubMed]
19. Shu, W.; Lu, M.M.; Zhang, Y.; Tucker, P.W.; Zhou, D.; Morrisey, E.E. Foxp2 and Foxp1 cooperatively regulate lung and esophagus development. *Development* **2007**, *134*, 1991–2000. [CrossRef] [PubMed]
20. Banham, A.H.; Beasley, N.; Campo, E.; Fernandez, P.L.; Fidler, C.; Gatter, K.; Jones, M.; Mason, D.Y.; Prime, J.E.; Trougouboff, P.; et al. The FOXP1 winged helix transcription factor is a novel candidate tumor suppressor gene on chromosome 3p. *Cancer Res.* **2001**, *61*, 8820–8829.
21. Shoemaker, M.B.; Husser, D.; Roselli, C.; Al Jazairi, M.; Chrispin, J.; Kuhne, M.; Neumann, B.; Knight, S.; Sun, H.; Mohanty, S.; et al. Genetic Susceptibility for Atrial Fibrillation in Patients Undergoing Atrial Fibrillation Ablation. *Circ. Arrhythm. Electrophysiol.* **2020**, *13*, e007676. [CrossRef]
22. Wang, Z.; Dolle, P.; Cardoso, W.V.; Niederreither, K. Retinoic acid regulates morphogenesis and patterning of posterior foregut derivatives. *Dev. Biol.* **2006**, *297*, 433–445. [CrossRef]
23. Chen, F.; Desai, T.J.; Qian, J.; Niederreither, K.; Lu, J.; Cardoso, W.V. Inhibition of Tgf beta signaling by endogenous retinoic acid is essential for primary lung bud induction. *Development* **2007**, *134*, 2969–2979. [CrossRef]
24. Lind, A.; Siersema, P.D.; Kusters, J.G.; Konijn, T.; Mebius, R.E.; Koenderman, L. The Microenvironment in Barrett's Esophagus Tissue Is Characterized by High FOXP3 and RALDH2 Levels. *Front. Immunol.* **2018**, *9*, 1375. [CrossRef] [PubMed]
25. Domyan, E.T.; Ferretti, E.; Throckmorton, K.; Mishina, Y.; Nicolis, S.K.; Sun, X. Signaling through BMP receptors promotes respiratory identity in the foregut via repression of Sox2. *Development* **2011**, *138*, 971–981. [CrossRef] [PubMed]
26. Castillo, D.; Puig, S.; Iglesias, M.; Seoane, A.; de Bolos, C.; Munitiz, V.; Parrilla, P.; Comerma, L.; Poulsom, R.; Krishnadath, K.K.; et al. Activation of the BMP4 pathway and early expression of CDX2 characterize non-specialized columnar metaplasia in a human model of Barrett's esophagus. *J. Gastrointest. Surg.* **2012**, *16*, 227–237. [CrossRef] [PubMed]
27. Greten, F.R.; Grivennikov, S.I. Inflammation and Cancer: Triggers, Mechanisms, and Consequences. *Immunity* **2019**, *51*, 27–41. [CrossRef]
28. Kaur, B.S.; Triadafilopoulos, G. Acid- and bile-induced PGE(2) release and hyperproliferation in Barrett's esophagus are COX-2 and PKC-epsilon dependent. *Am. J. Physiol. Gastrointest. Liver Physiol.* **2002**, *283*, G327–G334. [CrossRef]
29. Souza, R.F.; Shewmake, K.; Terada, L.S.; Spechler, S.J. Acid exposure activates the mitogen-activated protein kinase pathways in Barrett's esophagus. *Gastroenterology* **2002**, *122*, 299–307. [CrossRef]
30. Abdel-Latif, M.M.; O'Riordan, J.; Windle, H.J.; Carton, E.; Ravi, N.; Kelleher, D.; Reynolds, J.V. NF-kappaB activation in esophageal adenocarcinoma: Relationship to Barrett's metaplasia, survival, and response to neoadjuvant chemoradiotherapy. *Ann. Surg.* **2004**, *239*, 491–500. [CrossRef]
31. Morgan, C.; Alazawi, W.; Sirieix, P.; Freeman, T.; Coleman, N.; Fitzgerald, R. In vitro acid exposure has a differential effect on apoptotic and proliferative pathways in a Barrett's adenocarcinoma cell line. *Am. J. Gastroenterol.* **2004**, *99*, 218–224. [CrossRef]
32. Quante, M.; Bhagat, G.; Abrams, J.A.; Marache, F.; Good, P.; Lee, M.D.; Lee, Y.; Friedman, R.; Asfaha, S.; Dubeykovskaya, Z.; et al. Bile acid and inflammation activate gastric cardia stem cells in a mouse model of Barrett-like metaplasia. *Cancer Cell* **2012**, *21*, 36–51. [CrossRef]

33. El-Serag, H.B.; Aguirre, T.V.; Davis, S.; Kuebeler, M.; Bhattacharyya, A.; Sampliner, R.E. Proton pump inhibitors are associated with reduced incidence of dysplasia in Barrett's esophagus. *Am. J. Gastroenterol.* **2004**, *99*, 1877–1883. [CrossRef]
34. Kastelein, F.; Spaander, M.C.; Biermann, K.; Steyerberg, E.W.; Kuipers, E.J.; Bruno, M.J. Nonsteroidal anti-inflammatory drugs and statins have chemopreventative effects in patients with Barrett's esophagus. *Gastroenterology* **2011**, *141*, 2000–2008; quiz e2013–e2014. [CrossRef]
35. Jankowski, J.A.Z.; de Caestecker, J.; Love, S.B.; Reilly, G.; Watson, P.; Sanders, S.; Ang, Y.; Morris, D.; Bhandari, P.; Brooks, C.; et al. Esomeprazole and aspirin in Barrett's oesophagus (AspECT): A randomised factorial trial. *Lancet* **2018**, *392*, 400–408. [CrossRef]
36. Dvorak, K.; Payne, C.M.; Chavarria, M.; Ramsey, L.; Dvorakova, B.; Bernstein, H.; Holubec, H.; Sampliner, R.E.; Guy, N.; Condon, A.; et al. Bile acids in combination with low pH induce oxidative stress and oxidative DNA damage: Relevance to the pathogenesis of Barrett's oesophagus. *Gut* **2007**, *56*, 763–771. [CrossRef] [PubMed]
37. De Gottardi, A.; Hadengue, A.; Dumonceau, J.M. Retinoids, bile acids and PPARs in Barrett's oesophagus. *Gut* **2008**, *57*, 137. [CrossRef]
38. Lam-Himlin, D.M.; Daniels, J.A.; Gayyed, M.F.; Dong, J.; Maitra, A.; Pan, D.; Montgomery, E.A.; Anders, R.A. The hippo pathway in human upper gastrointestinal dysplasia and carcinoma: A novel oncogenic pathway. *Int. J. Gastrointest. Cancer* **2006**, *37*, 103–109. [CrossRef] [PubMed]
39. Zhao, B.; Wei, X.; Li, W.; Udan, R.S.; Yang, Q.; Kim, J.; Xie, J.; Ikenoue, T.; Yu, J.; Li, L.; et al. Inactivation of YAP oncoprotein by the Hippo pathway is involved in cell contact inhibition and tissue growth control. *Genes Dev.* **2007**, *21*, 2747–2761. [CrossRef]
40. Dietrich, C.; Kaina, B. The aryl hydrocarbon receptor (AhR) in the regulation of cell-cell contact and tumor growth. *Carcinogenesis* **2010**, *31*, 1319–1328. [CrossRef]
41. Zhao, B.; Li, L.; Wang, L.; Wang, C.Y.; Yu, J.; Guan, K.L. Cell detachment activates the Hippo pathway via cytoskeleton reorganization to induce anoikis. *Genes Dev.* **2012**, *26*, 54–68. [CrossRef]
42. Koivusalo, A.I.; Rintala, R.J.; Pakarinen, M.P. Outcomes of fundoplication in oesophageal atresia associated gastrooesophageal reflux disease. *J. Pediatric Surg.* **2018**, *53*, 230–233. [CrossRef] [PubMed]
43. Van Wijk, M.; Knuppe, F.; Omari, T.; de Jong, J.; Benninga, M. Evaluation of gastroesophageal function and mechanisms underlying gastroesophageal reflux in infants and adults born with esophageal atresia. *J. Pediatric Surg.* **2013**, *48*, 2496–2505. [CrossRef] [PubMed]
44. Tugay, M.; Yildiz, F.; Utkan, T.; Ulak, G.; Gacar, N.; Erden, F. Impaired esophageal reactivity in adriamycin-induced rat esophageal atresia: An in vitro study. *J. Pediatric Surg.* **2001**, *36*, 1569–1573. [CrossRef] [PubMed]
45. Kauer, W.K.; Peters, J.H.; DeMeester, T.R.; Feussner, H.; Ireland, A.P.; Stein, H.J.; Siewert, R.J. Composition and concentration of bile acid reflux into the esophagus of patients with gastroesophageal reflux disease. *Surgery* **1997**, *122*, 874–881. [CrossRef]
46. Zaninotto, G.; Di Mario, F.; Costantini, M.; Baffa, R.; Germana, B.; Dal Santo, P.L.; Rugge, M.; Bolzan, M.; Naccarato, R.; Ancona, E. Oesophagitis and pH of refluxate: An experimental and clinical study. *Br. J. Surg.* **1992**, *79*, 161–164. [CrossRef]
47. Leedham, S.J.; Preston, S.L.; McDonald, S.A.; Elia, G.; Bhandari, P.; Poller, D.; Harrison, R.; Novelli, M.R.; Jankowski, J.A.; Wright, N.A. Individual crypt genetic heterogeneity and the origin of metaplastic glandular epithelium in human Barrett's oesophagus. *Gut* **2008**, *57*, 1041–1048. [CrossRef]
48. Martinez, P.; Mallo, D.; Paulson, T.G.; Li, X.; Sanchez, C.A.; Reid, B.J.; Graham, T.A.; Kuhner, M.K.; Maley, C.C. Evolution of Barrett's esophagus through space and time at single-crypt and whole-biopsy levels. *Nat. Commun.* **2018**, *9*, 794. [CrossRef]
49. Jego, M.; Volant, A.; Faycal, J.; Doucet, L.; Andlauer, E.; Delalande, A.H.; Cholet, F.; Nousbaum, J.B.; Gouerou, H.; Robaszkiewicz, M. Prevalence and topography of intestinal metaplasia in columnar lined esophagus. *Gastroenterol. Clin. Biol.* **2007**, *31*, 601–606. [CrossRef]
50. Yamamoto, Y.; Wang, X.; Bertrand, D.; Kern, F.; Zhang, T.; Duleba, M.; Srivastava, S.; Khor, C.C.; Hu, Y.; Wilson, L.H.; et al. Mutational spectrum of Barrett's stem cells suggests paths to initiation of a precancerous lesion. *Nat. Commun.* **2016**, *7*, 10380. [CrossRef]
51. Mourikis, T.P.; Benedetti, L.; Foxall, E.; Temelkovski, D.; Nulsen, J.; Perner, J.; Cereda, M.; Lagergren, J.; Howell, M.; Yau, C.; et al. Patient-specific cancer genes contribute to recurrently perturbed pathways and establish therapeutic vulnerabilities in esophageal adenocarcinoma. *Nat. Commun.* **2019**, *10*, 3101. [CrossRef] [PubMed]
52. Zhong, G.; Ahimaz, P.; Edwards, N.A.; Hagen, J.J.; Faure, C.; Kingma, P.; Middlesworth, W.; Khlevner, J.; Fiky, M.E.; Schindel, D.; et al. Identification and validation of novel candidate risk genes in endocytic vesicular trafficking associated with esophageal atresia and tracheoesophageal fistulas. *medRxiv* **2021**. [CrossRef]
53. Brosens, E.; Marsch, F.; de Jong, E.M.; Zaveri, H.P.; Hilger, A.C.; Choinitzki, V.G.; Holscher, A.; Hoffmann, P.; Herms, S.; Boemers, T.M.; et al. Copy number variations in 375 patients with oesophageal atresia and/or tracheoesophageal fistula. *Eur. J. Hum. Genet.* **2016**, *24*, 1715–1723. [CrossRef] [PubMed]
54. Iftikhar, S.Y.; James, P.D.; Steele, R.J.; Hardcastle, J.D.; Atkinson, M. Length of Barrett's oesophagus: An important factor in the development of dysplasia and adenocarcinoma. *Gut* **1992**, *33*, 1155–1158. [CrossRef] [PubMed]
55. Palles, C.; Chegwidden, L.; Li, X.; Findlay, J.M.; Farnham, G.; Castro Giner, F.; Peppelenbosch, M.P.; Kovac, M.; Adams, C.L.; Prenen, H.; et al. Polymorphisms near tbx5 and gdf7 are associated with increased risk for barrett's esophagus. *Gastroenterology* **2015**, *148*, 367–378. [CrossRef] [PubMed]

56. Levine, D.M.; Ek, W.E.; Zhang, R.; Liu, X.; Onstad, L.; Sather, C.; Lao-Sirieix, P.; Gammon, M.D.; Corley, D.A.; Shaheen, N.J.; et al. A genome-wide association study identifies new susceptibility loci for esophageal adenocarcinoma and barrett's esophagus. *Nat. Genet.* **2013**, *45*, 1487–1493. [CrossRef]
57. Casson, A.G.; Zheng, Z.; Evans, S.C.; Geldenhuys, L.; van Zanten, S.V.; Veugelers, P.J.; Porter, G.A.; Guernsey, D.L. Cyclin d1 polymorphism (g870a) and risk for esophageal adenocarcinoma. *Cancer* **2005**, *104*, 730–739. [CrossRef]
58. Ren, D.; Zheng, G.; Bream, S.; Tevebaugh, W.; Shaheen, N.J.; Chen, X. Single nucleotide polymorphisms of caudal type homeobox 1 and 2 are associated with barrett's esophagus. *Dig. Dis. Sci.* **2014**, *59*, 57–63. [CrossRef]
59. Su, Z.; Gay, L.J.; Strange, A.; Palles, C.; Band, G.; Whiteman, D.C.; Lescai, F.; Langford, C.; Nanji, M.; Edkins, S.; et al. Common variants at the mhc locus and at chromosome 16q24.1 predispose to barrett's esophagus. *Nat. Genet.* **2012**, *44*, 1131–1136. [CrossRef]
60. Becker, J.; May, A.; Gerges, C.; Anders, M.; Veits, L.; Weise, K.; Czamara, D.; Lyros, O.; Manner, H.; Terheggen, G.; et al. Supportive evidence for foxp1, barx1, and foxf1 as genetic risk loci for the development of esophageal adenocarcinoma. *Cancer Med.* **2015**, *4*, 1700–1704. [CrossRef]
61. Kala, Z.; Dolina, J.; Marek, F.; Izakovicova Holla, L. Polymorphisms of glutathione s-transferase m1, t1 and p1 in patients with reflux esophagitis and barrett's esophagus. *J. Hum. Genet.* **2007**, *52*, 527–534. [CrossRef] [PubMed]
62. Bull, L.M.; White, D.L.; Bray, M.; Nurgalieva, Z.; El-Serag, H.B. Phase i and ii enzyme polymorphisms as risk factors for barrett's esophagus and esophageal adenocarcinoma: A systematic review and meta-analysis. *Dis. Esophagus* **2009**, *22*, 571–587. [CrossRef] [PubMed]
63. McElholm, A.R.; McKnight, A.J.; Patterson, C.C.; Johnston, B.T.; Hardie, L.J.; Murray, L.J.; Finbar, G. A population-based study of igf axis polymorphisms and the esophageal inflammation, metaplasia, adenocarcinoma sequence. *Gastroenterology* **2010**, *139*, 204–212.e3. [CrossRef]
64. Moons, L.M.; Kusters, J.G.; van Delft, J.H.; Kuipers, E.J.; Gottschalk, R.; Geldof, H.; Bode, W.A.; Stoof, J.; van Vliet, A.H.; Ketelslegers, H.B.; et al. A pro-inflammatory genotype predisposes to barrett's esophagus. *Carcinogenesis* **2008**, *29*, 926–931. [CrossRef] [PubMed]
65. Buas, M.F.; He, Q.; Johnson, L.G.; Onstad, L.; Levine, D.M.; Thrift, A.P.; Gharahkhani, P.; Palles, C.; Lagergren, J.; Fitzgerald, R.C.; et al. Germline variation in inflammation-related pathways and risk of barrett's oesophagus and oesophageal adenocarcinoma. *Gut* **2017**, *66*, 1739–1747. [CrossRef]
66. Yan, C.; Ji, Y.; Huang, T.; Yu, F.; Gao, Y.; Gu, Y.; Qi, Q.; Du, J.; Dai, J.; Ma, H.; et al. An esophageal adenocarcinoma susceptibility locus at 9q22 also confers risk to esophageal squamous cell carcinoma by regulating the function of barx1. *Cancer Lett.* **2018**, *421*, 103–111. [CrossRef]
67. Abnet, C.C.; Freedman, N.D.; Hu, N.; Wang, Z.; Yu, K.; Shu, X.O.; Yuan, J.M.; Zheng, W.; Dawsey, S.M.; Dong, L.M.; et al. A shared susceptibility locus in plce1 at 10q23 for gastric adenocarcinoma and esophageal squamous cell carcinoma. *Nat. Genet.* **2010**, *42*, 764–767. [CrossRef]
68. Dura, P.; van Veen, E.M.; Salomon, J.; te Morsche, R.H.; Roelofs, H.M.; Kristinsson, J.O.; Wobbes, T.; Witteman, B.J.; Tan, A.C.; Drenth, J.P.; et al. Barrett associated mhc and foxf1 variants also increase esophageal carcinoma risk. *Int. J. Cancer* **2013**, *133*, 1751–1755. [CrossRef]
69. Wideroff, L.; Vaughan, T.L.; Farin, F.M.; Gammon, M.D.; Risch, H.; Stanford, J.L.; Chow, W.H. Gst, nat1, cyp1a1 polymorphisms and risk of esophageal and gastric adenocarcinomas. *Cancer Detect. Prev.* **2007**, *31*, 233–236. [CrossRef]
70. Gross, R.E. *The Surgery of Infancy and Childhood: Its Principles and Techniques*; W.B. Saunders Co.: Philadelphia, PA, USA, 1953.
71. Lundell, L.R.; Dent, J.; Bennett, J.R.; Blum, A.L.; Armstrong, D.; Galmiche, J.P.; Johnson, F.; Hongo, M.; Richter, J.E.; Spechler, S.J.; et al. Endoscopic assessment of oesophagitis: Clinical and functional correlates and further validation of the los angeles classification. *Gut* **1999**, *45*, 172–180. [CrossRef]
72. Rasanen, K.; Itkonen, O.; Koistinen, H.; Stenman, U.H. Emerging roles of spink1 in cancer. *Clin. Chem.* **2016**, *62*, 449–457. [CrossRef] [PubMed]
73. Jennings, R.E.; Berry, A.A.; Kirkwood-Wilson, R.; Roberts, N.A.; Hearn, T.; Salisbury, R.J.; Blaylock, J.; Piper Hanley, K.; Hanley, N.A. Development of the human pancreas from foregut to endocrine commitment. *Diabetes* **2013**, *62*, 3514–3522. [CrossRef] [PubMed]
74. Gene Expression for Spink1 (ensg00000164266.10). Available online: https://gtexportal.org/home/gene/SPINK1 (accessed on 15 October 2021).
75. Alexander, D.H.; Lange, K. Enhancements to the admixture algorithm for individual ancestry estimation. *BMC Bioinform.* **2011**, *12*, 246. [CrossRef] [PubMed]
76. Wray, N.R.; Goddard, M.E.; Visscher, P.M. Prediction of individual genetic risk to disease from genome-wide association studies. *Genome Res.* **2007**, *17*, 1520–1528. [CrossRef]
77. Clc Genomics Workbench User Manual. Available online: https://digitalinsights.qiagen.com/technical-support/manuals/ (accessed on 15 October 2021).
78. Bus, P.; Siersema, P.D.; Verbeek, R.E.; van Baal, J.W. Upregulation of mirna-143, -145, -192, and -194 in esophageal epithelial cells upon acidic bile salt stimulation. *Dis. Esophagus* **2014**, *27*, 591–600. [CrossRef] [PubMed]

cancers

Article

Clinical, Molecular and Genetic Characteristics of Early Onset Gastric Cancer: Analysis of a Large Multicenter Study

Anna Pocurull [1], Cristina Herrera-Pariente [1], Sabela Carballal [1], Joan Llach [1], Ariadna Sánchez [1], Laura Carot [2], Josep María Botargues [3], Miriam Cuatrecasas [4], Teresa Ocaña [1], Francesc Balaguer [1], Luis Bujanda [5] and Leticia Moreira [1,*]

[1] Gastroenterology Department, Hospital Clínic de Barcelona, Centro de Investigación Biomédica en Red de Enfermedades Hepáticas y Digestivas (CIBERehd), Institut d'Investigacions Biomediques August Pi i Sunyer (IDIBAPS), Universitat de Barcelona, 08036 Barcelona, Spain; pocurull@clinic.cat (A.P.); cristina.herrera@ciberehd.org (C.H.-P.); carballal@clinic.cat (S.C.); jllachr@clinic.cat (J.L.); asanchezg@clinic.cat (A.S.); mocana@clinic.cat (T.O.); fprunes@clinic.cat (F.B.)

[2] Gastroenterology Department, Hospital del Mar, 08003 Barcelona, Spain; lcarot@psmar.cat

[3] Gastroenterology Department, Hospital Universitari de Bellvitge, 08097 Barcelona, Spain; jbotargues@bellvitgehospital.cat

[4] Department Pathology, Hospital Clínic de Barcelona, Centro de Investigación Biomédica en Red de Enfermedades Hepáticas y Digestivas (CIBERehd), Institut d'Investigacions Biomediques August Pi i Sunyer (IDIBAPS), Universitat de Barcelona, 08036 Barcelona, Spain; mcuatrec@clinic.cat

[5] Gastroenterology Department, Biodonostia Health Research Institute, Centro de Investigación Biomédica en Red de Enfermedades Hepáticas y Digestivas (CIBERehd), Universidad del País Vasco (UPV/EHU), 20014 San Sebastián, Spain; LUIS.BUJANDAFERNANDEZDEPIEROLA@osakidetza.eus

* Correspondence: lmoreira@clinic.cat; Tel.: +34-93-227-5739

Citation: Pocurull, A.; Herrera-Pariente, C.; Carballal, S.; Llach, J.; Sánchez, A.; Carot, L.; Botargues, J.M.; Cuatrecasas, M.; Ocaña, T.; Balaguer, F.; et al. Clinical, Molecular and Genetic Characteristics of Early Onset Gastric Cancer: Analysis of a Large Multicenter Study. *Cancers* **2021**, *13*, 3132. https://doi.org/10.3390/cancers13133132

Academic Editor: Irit Ben Aharon

Received: 17 March 2021
Accepted: 18 June 2021
Published: 23 June 2021

Simple Summary: Gastric cancer is one of the most common cancers worldwide, showing high mortality rates. A small portion of gastric cancer patients, known as early onset gastric cancer (EOGC) patients, develop the disease before age 50, and their characteristics are poorly described. Thus, our main objective was to describe the clinical, molecular, and genetic characteristics of EOGC in a large multicenter cohort of patients. We were able to identify that most EOGC cases have similar characteristics: diagnosed at advanced stage, diffuse type, and infrequent DNA mismatch repair somatic deficiency. Although familial aggregation of gastric cancer was uncommon, a germline genetic mutation was identified in 25% of the patients tested. Our results show that EOGC has a marked genetic heterogeneity. Thus, it is essential to consider familial history of tumors, not only GC, in order to select adequate patients to perform a suitable genetic counseling and enhance the emerging use of multigene panels.

Abstract: Gastric adenocarcinoma (GC) is a common tumor with high morbidity and mortality. Only 7% of patients with GC are diagnosed before age 50 (early onset gastric cancer (EOGC)), and their characteristics have been poorly described. We aimed to describe clinical, molecular, and genetic characteristics of EOGC. A total of 309 patients with EOGC were retrospectively studied in four Spanish centers. Personal information, family history, and tumor information were registered. Germinal genetic analysis was performed in patients who met current criteria of a hereditary syndrome at the time of diagnosis. The median age at diagnosis was 44 years. The majority (73.3%) of tumors were diffuse, and 78.3% were diagnosed in an advanced stage. Familial aggregation of GC was present in 18/117 (15.4%) cases, and 5/117 (4.3%) met criteria for familial GC. MMR-IHC was performed in 126/309 (40.7%) tumors: 4/126 (3.1%) had loss of expression in MLH1/PMS2, without an associated germline mutation. Sixteen germline genetic analyses were performed, detecting a pathogenic variant in four (25%) cases: one in *BRCA2*, one in *TP53*, and two in *CDH1*. Most EOGC are diffuse and diagnosed in an advanced stage. In these patients, DNA MMR system deficiency is uncommon. Although familial aggregation was observed in only 15% of cases, a germline mutation was found in 25% of patients tested with clinical criteria. This demonstrates that EOGC has a marked genetic heterogeneity, reinforcing the importance of an accurate genetic counseling and enhancing the emerging use of multigene panels.

Keywords: gastric cancer; early onset cancer; DNA mismatch repair; hereditary cancer; familial cancer

1. Introduction

Gastric cancer (GC) is the fifth most common and the third most deadly cancer in the world [1], representing a worldwide health problem [2]. The average age at diagnosis is 60 years, only 7% occur before age 50 and 2% before age 40 [3]. The etiology of GC is multifactorial, with *Helicobacter pylori*, diet factors, and tobacco being the main environmental agents implicated in its pathogenesis [4]. Although most GC cases are sporadic, a familial aggregation is observed in approximately 10% of cases, with an underlying genetic cause identified in up to 5% of all GC [5]. Familial characteristics that suggest a hereditary predisposition include the existence of several affected family members, an autosomal dominant pattern of inheritance, disease presentation at young ages, and association with other extra-gastric neoplasms [6].

In terms of these assumptions, there are mainly three clinical situations where familial predisposition to GC may be found [7]. First, hereditary syndromes with higher risk for GC, including two entities: gastric adenocarcinoma and proximal polyposis of the stomach (GAPPS) [8], and the most common inherited GC syndrome, hereditary diffuse gastric cancer (HDGC). It is characterized by two or more cases of GC at any age in first or second relatives, with at least one confirmed diffuse gastric cancer (DGC); or personal history of DGC before the age of 50; or personal or family history (first- or second-degree relatives) of DGC and lobular breast cancer, with one being diagnosed before 70 years [9]. This syndrome is mainly caused by *CDH1* germline mutations, which encode the tumor suppressor protein E-cadherin. However, during the last years, another gene, *CTNNA1*, has also been identified in HDGC families [10].

Other clinical situations are hereditary syndromes with higher risk for GC and other tumors, including Lynch syndrome and, less commonly, familial adenomatous polyposis (FAP), Peutz–Jeghers syndrome, juvenile polyposis, Li–Fraumeni syndrome, and Cowden syndrome [11], with germline mutations in different genes. The last clinical situation is familial intestinal gastric cancer (FIGC) characterized by familial aggregation of intestinal GCs without an identified inherited cause. FIGC is defined as two or more cases of GC in first-degree (FDR) or second-degree relatives (SDR), with at least one confirmed case of intestinal histology in someone younger than 50 years, or three or more confirmed cases of intestinal GC in FDR or SDR, regardless of age [12,13].

In spite of this, the genetic cause is not identified in a high percentage of GC patients [14,15]. This fact is especially important in early-onset gastric cancer cases (EOGC) because 90% of these young patients do not have a family history, hampering identification and early diagnosis [16]. The remaining 10% of EOGC cases that have a family history are explained by the previously mentioned hereditary syndromes.

The definition of early onset gastric cancer (EOGC) varies across studies, but one of the most accepted definitions includes those diagnosed at the age of 50 or younger. Although the incidence of GC is declining globally, EOGC is increasing [17]. In fact, a recent study has reported that nowadays EOGC comprises up to 30% of all cases of GC in the United States [18]. EOGC has been associated with some clinical and pathological characteristics, such as predomination of diffuse histology and infrequent association with intestinal metaplasia [19,20]. Moreover, EOGC is usually diagnosed in an advanced stage, associated with a high mortality. However, clinical and molecular features of EOGC have been poorly described [16].

Identification of individuals at high risk of GC allows us to establish preventive measures, early diagnosis, and personalized treatments; thus, we aimed to describe the clinical, molecular, and genetic characteristics of EOGC ($\leq$50 years) in order to identify high-risk forms of GC.

2. Materials and Methods

2.1. Study Population

Patients with GC diagnosed before 51 years old were retrospectively studied at four centers in Spain between 1999 and 2018. The study was approved by the Institutional Review Board (or Ethics Committee) of Hospital Clínic in Barcelona (register number 2015/0153, date of approval 22/04/2015).

Clinical and demographic data were evaluated through electronic clinical reports, including environmental risks factors such as tobacco consumption, alcohol intake, and *Helicobacter pylori* infection.

Personal and family history of GC and other tumors related with hereditary syndromes (i.e., HDGC, Peutz–Jeghers syndrome, Lynch syndrome, familial adenomatous polyposis, HBOC, juvenile polyposis, and Li–Fraumeni syndrome) were registered (including FDRs and SDRs). Patients who met criteria of familial GC were also identified.

2.2. Tumor Characteristics

The tumor characteristics were analyzed by histological report. The location, diagnostic stage (TNM), histologic features (intestinal, diffuse, or mixed), and grade of differentiation were considered.

Tumor mismatch repair (MMR) deficiency was evaluated by immunostaining including analysis of MLH1, MSH2, MSH6, and PMS2 protein expression, as previously described [21].

2.3. Germline Genetic Analysis

Germline genetic testing was performed on genomic DNA isolated from peripheral blood leukocytes by both multiple ligation probe amplification analysis and direct sequencing. The analysis was performed through a commercial multigene panel (Trusight Cancer v1, Illumina Inc., San Diego, CA, USA) involving the most frequent genes related to GC germline predisposition (*MLH1, MSH2, MSH6, PMS2, CDH1, EPCAM, BRCA1, BRCA2, PALB2, TP53, APC, MUTYH, STK11, SMAD4/BMPR1A, PTEN*).

The genetic test was performed in patients with available germline DNA who fulfilled the diagnostic criteria of a hereditary syndrome related to GC at the time of diagnosis or in whom the tumoral analysis of DNA mismatch repair proteins was altered (i.e., loss of protein expression of MLH1, MSH2, MSH6, or PMS2) [13].

2.4. Statistical Analysis

All data were analyzed using the 22.0 SPSS software package (IBM SPSS Statistics for Window, Version 22.0. Armonk, NY: IBM Corp.).

Baseline characteristics were described in percentages for categorical data, using median, range, and interquartile range (IQR). When information was missing, the denominator was accordingly to patients with available data. Univariate binary logistic regression was performed for selection of variables associated with the diagnosis of a hereditary cancer. For multivariable logistic regression analyses, only candidate variables with p-values of $\leq$0.05 on univariate analysis were used in the final multivariate model. Odds ratios (ORs) with 95% confidence intervals (CIs) were included to quantify the magnitude of the association.

3. Results

3.1. General Characteristics

Three hundred and nine patients with EOGC were included. Clinico-pathological features of patients included in the study are summarized in Table 1.

Table 1. Clinical characteristics of the patients (N = 309).

	EOGC
Age (years) at diagnosis; median (IQR)	44 (40–48)
Gender, number (%): Man	191 (61.8)
Environmental risk factors; number (%)	
Helicobacter pylori infection	24/82 (29.3)
Smokers	77/169 (45.6)
Moderate or high alcohol consumers	21/105 (20)
Personal history of extra-gastric cancer (*n* = 309); number (%)	
Breast and/or ovarian	5 (1.6)
Lung	1 (0.3)
Thyroid	1 (0.3)
Hodgkin lymphoma	1 (0.3)
Cervix	1 (0.3)
Familial history of cancer (*n* = 117); number (%)	
GC	18 (15.4)
Colorectal	8 (6.8)
Ovarian and/or breast	15 (12.8)
Others	39 (33.3)
Criteria of familial GC	5 (4.3)

IQR, interquartile range; EOGC, early onset gastric cancer; GC, gastric cancer.

The median age at diagnosis was 44 years old (IQR 40-48, range 33), with a predominance of men, with 191 (61.8%) cases. Related to environmental risk factors, 77/169 (45%) were smokers and 21/105 (20%) had chronic alcohol consumption, whereas in 24/82 (29%) cases, a *Helicobacter pylori* infection was detected. Most of the patients were from Spain; however, 8/309 (2.6%) were from South America, 4/309 (1.3%) were from Asia, and 4/309 (1.3%) were South African.

Out of the 309 (2.9%) patients, 9 had previously developed other tumors, including breast and/or ovarian in 5/309 (1.6%), lung in 1/309 (0.3%), cervix in 1/309 (0.3%), thyroid in 1/309 (0.3%), and Hodgkin lymphoma in 1/309 (0.3%).

A total of 18/117 (15.4%) patients presented familial aggregation of GC (≥1 FDR or SDR affected), and 5/117 (4.3%) met criteria for FIGC. A total of 67/117 (57.2%) patients had family history of cancers related with a hereditary syndrome, mainly ovarian and/or breast in 15/117 (12.8%), followed by colorectal cancer in 8/117 (6.8%) cases and other types of tumors in 39/117 (34%). Detailed characteristics are described in Table 1.

3.2. Tumor Characteristics

The predominant tumor histology was diffuse, observed in 118/161 (73.3%) of the cases, and the signet ring cell subtype was detected in 38/118 (32%). Among patients with diffuse GC and *H. pylori* information available, in 8/45 (17.8%), the infection was present. Among patients with intestinal GC subtype with *H. pylori* status available, 3/14 (21.4%) were infected. No statistically significant differences regarding *H. pylori* infection and histology subtype were found ($p = 0.09$). According to the WHO classification (2019), the degree of differentiation only applies to the intestinal GC, and thus within this subtype, 9/43 (39.1%) were poorly differentiated tumors. Regarding tumor location, the most common sites were the body in 111/203 (55%) and antrum in 50/203 (25%). An advanced stage (III/IV) at diagnosis was present in 166/212 (78.3%) cases, and only in 44/212 (20.8%) was the diagnosis at an early stage (I/II).

In 122/205 (59.5%) cases, surgery with or without chemotherapy was performed; 67/205 (32.7%) patients were treated with chemotherapy (CT) +/− radiotherapy (RT) alone, and 16/205 (7.8%) patients did not receive a specific oncological treatment.

The immunohistochemistry of DNA mismatch repair proteins (MMR-IHC) was performed in 126 out of 309 (40.7%) tumors, and only 4/126 (3.1%) showed loss of protein expression, specifically MLH1/PMS2 (Figure 1). The tumor and MMR-IHC characteristics are shown in Table 2 and Figure 2, respectively.

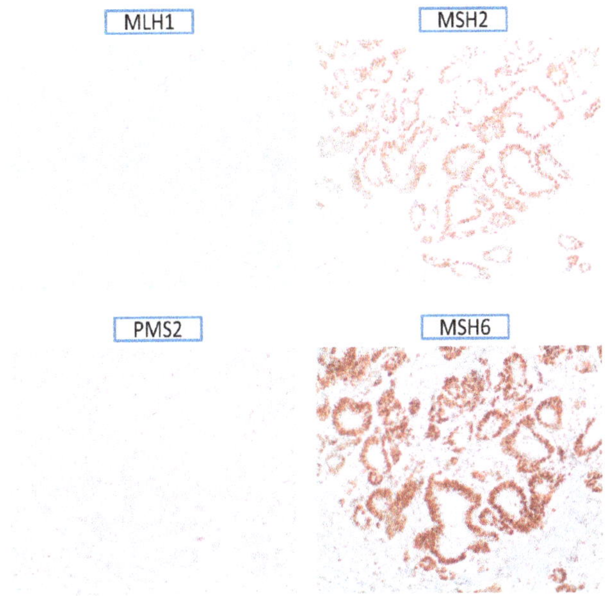

Figure 1. Immunohistochemistry of DNA mismatch repair proteins (MMR-IHC) in gastric cancer tissue loss of protein expression of MLH1 and PMS2, and normal protein expression of MSH2 and MSH6.

Table 2. Characteristics of the tumors (N = 309).

	EOGC
Gastric location (N = 203); number (%)	
Cardias	3 (1.5)
Fundus	18 (8.9)
Body	111 (54.7)
Antrum	50 (24.6)
Extensive	21 (10.3)
Stage (N = 212); number (%)	
I/II	44 (20.8)
III/IV	166 (78.3)
Histology (N = 161); number (%)	
(a) Diffuse	118 (73.3)
Signet ring cell subset	38/118 (32.2)
(b) Intestinal	30 (18.6)
(c) Mixed	13 (8.1)
Tumor differentiation grade (N = 23/43) number (%) *	
High grade (poorly differentiated)	9 (39.1)
Low-grade (well/moderately differentiated)	14 (60.9)
Treatment (N = 205) number (%)	
Surgery	40 (19.5)
Surgery + chemotherapy	58 (31.2)
Chemotherapy	64 (28.3)
Surgery + chemotherapy + radiotherapy	24 (11.7)
Chemotherapy + radiotherapy	3 (1.5)
Palliative	16 (7.8)

EOGC, early onset gastric cancer. * According to the WHO classification (2019), the degree of differentiation only applies to the intestinal type.

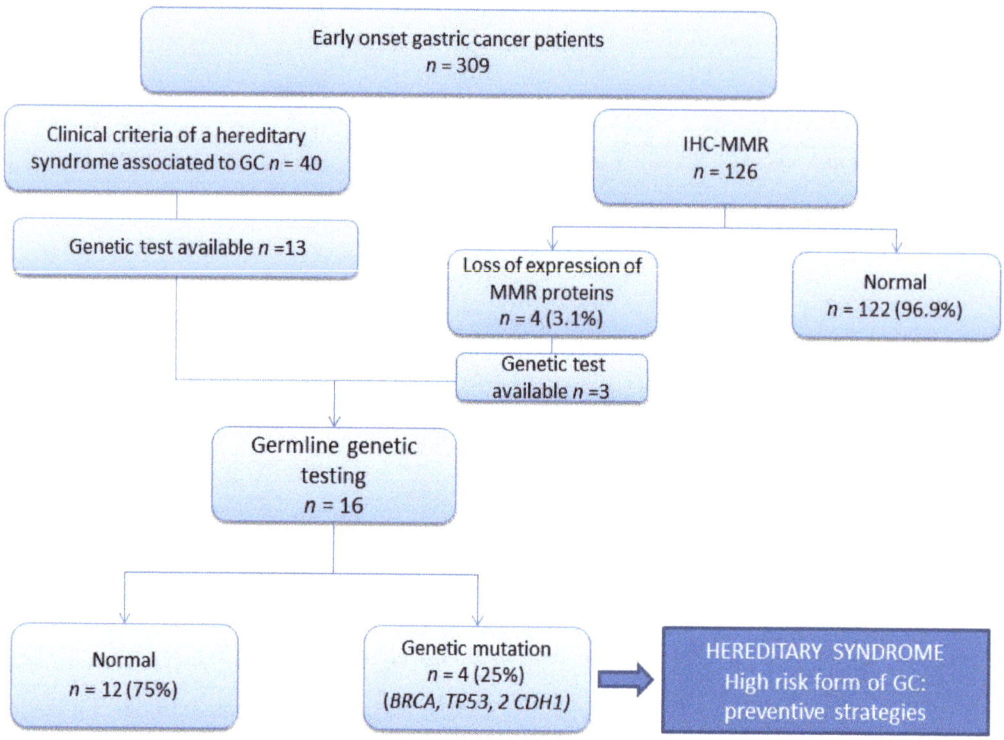

Figure 2. Scheme followed summarizing the results of IHC-MMR and germline genetic testing.

Regarding survival, with a median follow up of 7.6 years (IQR 17-38), a 5-year survival rate of 32.6% was observed, with a significant difference based on clinical stage (stage I–II 87% vs. stage III–IV 11.3%, $p = 0.0001$; Figure 3), and diffuse histology was associated with worse prognosis ($p = 0.019$); no differences in gender, age, family history of GC, *H. pylori* infection, smoking or alcohol consumption, tumor differentiation grade, or MMR-IHC were detected.

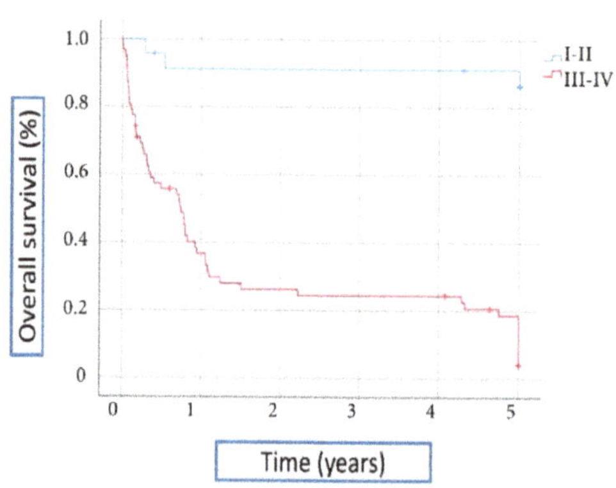

Figure 3. Overall survival rate based on diagnostic stage (stage I–II vs. stage III–IV). There is a significant difference in the 5-year survival rate: stage I-II 87% vs. stage III-IV 11.3%, $p = 0.0001$.

3.3. Germline Genetic Analysis

Genetic analysis was performed in the 16 patients with available germline DNA out of the 44 patients that fulfilled clinical criteria of germline testing. Among them, in 11/16 (68.7%) patients, the analysis was performed due to the fulfillment of criteria for HDGC, in 2/16 patients because they met criteria of HBOC; and in 3/16 cases, the genetic analysis was performed on the basis of a somatic loss of expression of MLH1/PMS2 proteins at IHC. In the remaining 28/44 patients, the analysis was not performed because the DNA was not available.

A germline genetic mutation was identified in 4/16 (25%) cases, one at 49 years old (with personal history of GC and breast cancer) and another three with GC younger than 41 years old. The mutated genes detected were *BRCA2* and *TP53*, and in two cases, *CDH1* (Table 3). Integrative Genomics Viewer was used for visualization of these variants; Figure S1 shows an example of two of these variants. None of those patients had tumors with loss of expression in DNA mismatch repair protein (MMR), and only one of them reported family history of GC. Within four patients with an altered MMR-IHC, the germline genetic analysis performed in three patients did not identify any pathogenic variant, and in the remaining patient, the germline analysis could not be performed because he died before the MMR-IHC was done.

Table 3. Characteristics in patients with hereditary syndromes.

Patient	Age	Gender	Tumor Characteristics	*H. pylori* Infection	Personal History of Other Tumors	Familial History of GC	Familial History of Other Tumors	MMR-IHC	Gene (Pathogenic Variant)
1	49	Woman	Intestinal hist. Stage II	Not available	Breast	No	Ovarian and breast	MMR+ (normal)	*BRCA2* (c.3166C>T; p.Gln1056 *; nonsense)
2	38	Man	Diffuse hist. Multifocal (plastic linitis) Stage IV	No	No	Yes	No	MMR+ (normal)	*CDH1* (c.2164+5G>C; splicing)
3	34	Man	Diffuse hist. Multifocal (plastic linitis) Stage IV	No	No	No	Breast and colorectal	MMR+ (normal)	*TP53* (c.365_366delTG; p.Val122fs; frameshift)
4	40	Man	Diffuse hist. Multifocal (plastic linitis) Stage IV	No	No	Yes	No	MMR+ (normal)	*CDH1* (c.187C>T; p.Arg63 *; nonsense)

hist., histology; GC, gastric cancer; MMR-IHC, immunohistochemistry of DNA mismatch repair proteins; MMR, tumor mismatch repair; *, stop codon

3.4. Factors Associated with High Risk of Gastric Cancer

Personal characteristics as well as family history were analyzed in order to identify risk factors of a hereditary GC syndrome (germline mutation identified). No statistically significant factor associated with the presence of a germline mutation was identified. However, family history of other neoplasms showed a trend towards statistical significance with $p = 0.057$.

4. Discussion

It is well known that EOGC has clinical and pathological differences with older onset GC [22], although their clinical and molecular characteristics have been poorly reported [16]. In the present study, we describe clinical, molecular, and genetic characteristics of 309 EOGC patients. Our study shows that most of EOGC are histologically diffuse (73%, in comparison with 32% in older patients), poorly differentiated, and diagnosed at an advanced stage, supporting what has been previously described in other studies [18–20,23]. Moreover, as the already reported trend of increasing rate in general population [24] of proximal GC over the distal location, we observed in our study that more than half of the tumors were located in the gastric body. This could be explained not only due to the diffuse histology, but also because the low incidence of *Helicobacter pylori* infection (less than 30%) and a high proportion (almost 60%) of patients with positive oncological family history, suggesting a different carcinogenesis process.

Although there are known characteristics of EOGC, there are some inconsistent data between studies, i.e., some studies reported a female predominance [23], while in others there was an increasing trend for males or without a significant difference between genders [25,26]. In the present study, a male predominance was identified.

On the basis of different features, our study attempted to deepen in the clinical and molecular characterization of EOGC with the ultimate goal of being able to identify high-risk individuals and establish preventive measures, early diagnosis, and personalized treatments.

Lynch syndrome (LS), one of the most common cancer hereditary syndromes, carries a cumulative risk of GC of 11–19% [27,28]. However, DNA mismatch repair deficiency is exceptional in GC [29]. In order to consider the diagnoses of possible LS, we analyzed the MMR system deficiency, observing that loss of protein expression was an infrequent event, and only 4/126 (3.1%) patients displayed it. The low incidence of MMR deficiency in EOGC is probably related to the high proportion of diffuse tumors, wherein MSI is less common [30] and also related to the fact that genomically stable tumors are usually diagnosed at an earlier age [31]. Moreover, in cases with MMR deficiency, we did not find a correlation with a germline mutation, suggesting a somatic loss origin due to hypermethylation of the *MLH1* promoter gene [32]. Thus, based upon our results, in this subgroup of patients, systematically analysis of MMR deficiency to rule out Lynch syndrome through IHC is likely not very useful.

Germline analysis was performed in 16 patients, representing 36% of patients who met clinical criteria for genetic testing according with the current guideline at that moment of the study (i.e., 2018–2019) [13]. Despite the fact that genetic testing could not be performed in the whole cohort of patients (due to loss of follow up or death), we found a germline mutation in 25% of tested patients. These mutations were located on *BRCA2*, *CDH1*, and *TP53* genes. Both patients with *CDH1* germline mutations displayed familial history of GC, while patients with *BRCA2* and *TP53* germline mutations showed familial history of ovarian and breast cancer, and breast and colorectal cancer, respectively. Thus, although familial aggregation of GC was present in only 15% of cases and the majority of patients with a germline mutation did not have familial aggregation of GC, family history of other tumors related with a hereditary syndrome was common. Therefore, this observation reinforces that, although family history of GC is poorly predictive, a very accurate medical personal and family history including any tumor type is mandatory in order to select the appropriate candidates for genetic testing.

Analyzing other studies, Tedaldi et al. focused on 96 patients that fulfilled different criteria such as HDGC criteria, suspected Lynch syndrome, familial aggregation, or patients with polyps and family history of GC. They sequenced 94 genes involved in cancer predisposition, identifying eight different *CDH1* pathogenic/likely patogenic mutations in nine different patients with DGC and a mean age of almost 40 years. Although they identified more variants in other genes, their carriers were patients with more than 50 years old, and therefore they cannot be considered as an early onset cohort [33]. Moreover, a Canadian cohort was studied using single-site and multi-gene panels. The authors identified mutations in *CDH1* and *BRCA2* in five and two patients with DGC before the age of 50, respectively. Comparing with the present study, similar results were obtained, although the *TP53* gene was not identified in this cohort [34]. However, a study performed by Vogelaar et al. did not identify any mutation neither in *CDH1* nor in *CTNNA1* in a cohort of 54 GC EOGC patients [35].

The relevance of the diagnosis of a hereditary syndrome is the opportunity to establish prevention and early diagnosis measures. For example, specifically in the context of *CDH1* mutation carriers, most asymptomatic individuals do not have macroscopic lesions on endoscopic examinations; however, intramucosal foci of gastric cancer, usually multiple, are observed in the surgical specimens. Therefore, it is recommended that one perform prophylactic total gastrectomy in carriers of a pathogenic variant who are older than 20 years [36,37]. Annual endoscopic screening is reserved for individuals who do not accept prophylactic gastrectomy, patients with a variant of uncertain significance, and patients in whom the germline mutation has not been identified. This point is reflected in our cohort, wherein the two patients with a *CDH1* mutation were diagnosed at stage IV (metastatic). In one case, two relatives were carriers of the mutation, with normal upper gastroscopy but with multiple focuses of diffuse adenocarcinoma, both in early stages (T1a and T1b, N0, M0). In the second patient, no additional *CDH1* mutation carriers were identified.

This study has some limitations, most of them because the data obtained were collected retrospectively, which implies potential inclusion biases and the difficulty to obtain the information of some variables: Epstein–Barr virus infection; associated gastritis; HER-2 or PDL1 status; and, as mentioned previously, germline testing, which was only evaluated in 16/44 (36%) of patients who met clinical criteria for genetic testing, and *CTNNA1* was not performed. Moreover, it is important to mention that *H. pylori* status was available in only 82 cases, being positive in 24 of them (29%); however, no difference regarding *H. pylori* prevalence among diffuse and intestinal histology was detected. In this context, other studies have been focused on the role of *H. pylori* infection and EOGC development, suggesting that it is important for tumor development [24], but to a lesser extent than in older GC patients [38]. Moreover, Rugge, et al. confirmed that *H. pylori* infection was significantly associated with both diffuse and intestinal histotypes [39]. In this sense, the low prevalence of *H. pylori* in our cohort, and the lack of differences in *H. pylori* status between histotypes, suggest that the implication of this infection in the carcinogenesis process is less relevant in EOGC, although prospective and large cohorts are needed to deepen in this observation.

The main strength of our study is that it describes not only clinical but also histological and molecular data of a large cohort of more than 300 patients with EOGC, although we are aware of the limitations associated to an observational and retrospective study.

5. Conclusions

Our results show that that most early onset GC cases are diagnosed in advanced stage, have diffuse histology, and have infrequent DNA mismatch repair somatic deficiency. Moreover, early onset GC has a marked genetic heterogeneity. Thus, it is essential to consider familial history of tumors, not only GC, but also and more importantly other tumors related with hereditary syndromes (such as colorectal, breast, and ovarian cancer), in order to select adequate patients to perform a suitable genetic counseling and enhance the emerging use of multigene panels.

Supplementary Materials: The following are available online at https://www.mdpi.com/article/10.3390/cancers13133132/s1, Figure S1: Example of visualization using Integrative Genomics Viewer of two of the germline variants identified. (a) *CDH1* c.2164+5G>C (splicing variant); (b) *TP53* c.365_366delTG (p.Val122fs; frameshift variant).

Author Contributions: Conceptualization, L.M., A.P., S.C., C.H.-P.; methodology, L.M., A.P., S.C., C.H.-P.; formal analysis, A.P., C.H.-P., L.M.; investigation, A.P., S.C., L.C., J.M.B., M.C., T.O., L.M., C.H.-P., A.S., J.L., L.B.; data curation, L.M., A.P., S.C., C.H.-P., F.B., L.B.; writing—original draft preparation, L.M., A.P., C.H.-P.; writing—review and editing, A.P., S.C., L.B., F.B., L.M.; supervision, L.M.; project administration, L.M.; funding acquisition, L.M. All authors have read and agreed to the published version of the manuscript.

Funding: This work was supported by CIBEREHD (grant 18PI03). CIBEREHD is funded by the Instituto de Salud Carlos III.

Institutional Review Board Statement: The study was conducted according to the guidelines of the Declaration of Helsinki and approved by the Institutional Review Board (or Ethics Committee) of Hospital Clínic in Barcelona (register number 2015/0153, date of approval 22/04/2015).

Informed Consent Statement: Informed consent was obtained from all subjects involved in the study.

Data Availability Statement: The data presented in this study are available on request from the corresponding author.

Conflicts of Interest: The authors declare no conflict of interest.

References

1. Ferlay, J.; Soerjomataram, I.; Dikshit, R.; Eser, S.; Mathers, C.; Rebelo, M.; Parkin, D.M.; Forman, D.; Bray, F. Cancer Incidence and Mortality Worldwide: Sources, Methods and Major Patterns in GLOBOCAN 2012. *Int. J. Cancer* **2015**, *136*, E359–E386. [CrossRef]
2. Gerdner, L.A. Individualized Music for Dementia: Evolution and Application of Evidence-Based Protocol. *World J. Psychiatry* **2012**, *2*, 26. [CrossRef] [PubMed]
3. Kluijt, I.; Sijmons, R.H.; Hoogerbrugge, N.; Plukker, J.T.; De Jong, D.; Van Krieken, J.H.; Van Hillegersberg, R.; Ligtenberg, M.; Bleiker, E.; Cats, A. Familial Gastric Cancer: Guidelines for Diagnosis, Treatment and Periodic Surveillance. *Fam. Cancer* **2012**, *11*, 363–369. [CrossRef] [PubMed]
4. Yusefi, A.R.; Lankarani, K.B.; Bastani, P.; Radinmanesh, M.; Kavosi, Z. Risk Factors for Gastric Cancer: A Systematic Review. *Asian Pac. J. Cancer Prev.* **2018**, *19*, 591–603. [PubMed]
5. Lott, P.C.; Carvajal-Carmona, L.G. Resolving Gastric Cancer Aetiology: An Update in Genetic Predisposition. *Lancet Gastroenterol. Hepatol.* **2018**, *3*, 874–883. [CrossRef]
6. Llach, J.; Moreno, L.; Sánchez, A.; Herrera-Pariente, C.; Ocaña, T.; Cuatrecasas, M.; Rivero-Sánchez, L.; Moreira, R.; Díaz, M.; Jung, G.; et al. Genetic Counseling for Hereditary Gastric and Pancreatic Cancer in High-Risk Gastrointestinal Cancer Clinics: An Effective Strategy. *Cancers* **2020**, *12*, 2386. [CrossRef]
7. Corso, G.; Roncalli, F.; Marrelli, D.; Carneiro, F.; Roviello, F. History, Pathogenesis, and Management of Familial Gastric Cancer: Original Study of John XXIII's Family. *Biomed. Res. Int.* **2013**, *2013*, 385132. [CrossRef]
8. Carneiro, F. Hereditary Gastric Cancer. *Pathologe* **2012**, *33* (Suppl. 2), 231–234. [CrossRef]
9. Blair, V.R.; McLeod, M.; Carneiro, F.; Coit, D.G.; D'Addario, J.L.; van Dieren, J.M.; Harris, K.L.; Hoogerbrugge, N.; Oliveira, C.; van der Post, R.S.; et al. Hereditary Diffuse Gastric Cancer: Updated Clinical Practice Guidelines. *Lancet Oncol.* **2020**, *21*, e386–e397. [CrossRef]
10. Majewski, I.J.; Kluijt, I.; Cats, A.; Scerri, T.S.; De Jong, D.; Kluin, R.J.C.; Hansford, S.; Hogervorst, F.B.L.; Bosma, A.J.; Hofland, I.; et al. An α-E-Catenin (CTNNA1) Mutation in Hereditary Diffuse Gastric Cancer. *J. Pathol.* **2013**, *229*, 621–629. [CrossRef]
11. Leoz, M.L.; Sánchez, A.; Carballal, S.; Ruano, L.; Ocaña, T.; Pellisé, M.; Castells, A.; Balaguer, F.; Moreira, L. Síndromes de Predisposición a Cáncer Gástrico y Cáncer Pancreático. *Gastroenterol. Hepatol.* **2016**, *39*, 481–493. [CrossRef]
12. Oliveira, C.; Pinheiro, H.; Figueiredo, J.; Seruca, R.; Carneiro, F. Familial Gastric Cancer: Genetic Susceptibility, Pathology, and Implications for Management. *Lancet Oncol.* **2015**, *16*, e60–e70. [CrossRef]
13. Stjepanovic, N.; Moreira, L.; Carneiro, F.; Balaguer, F.; Cervantes, A.; Balmaña, J.; Martinelli, E. Hereditary Gastrointestinal Cancers: ESMO Clinical Practice Guidelines for Diagnosis, Treatment and Follow-Up. *Ann. Oncol.* **2019**, *30*, 1558–1571. [CrossRef] [PubMed]
14. Hansford, S.; Kaurah, P.; Li-Chang, H.; Woo, M.; Senz, J.; Pinheiro, H.; Schrader, K.A.; Schaeffer, D.F.; Shumansky, K.; Zogopoulos, G.; et al. Hereditary Diffuse Gastric Cancer Syndrome: CDH1 Mutations and Beyond. *JAMA Oncol.* **2015**, *1*, 23–32. [CrossRef] [PubMed]
15. Herrera-Pariente, C.; Capó-García, R.; Díaz-Gay, M.; Carballal, S.; Muñoz, J.; Llach, J.; Sánchez, A.; Bonjoch, L.; Arnau-Collell, C.; de Lima, Y.S.; et al. Identification of New Genes Involved in Germline Predisposition to Early-Onset Gastric Cancer. *Int. J. Mol. Sci.* **2021**, *22*, 1310. [CrossRef]
16. Milne, A.N.; Offerhaus, G.J.A. Early-Onset Gastric Cancer: Learning Lessons from the Young. *World J. Gastrointest. Oncol.* **2010**, *2*, 59–64. [CrossRef] [PubMed]
17. Anderson, W.F.; Camargo, M.C.; Fraumeni, J.F.; Correa, P.; Rosenberg, P.S.; Rabkin, C.S. Age-Specific Trends in Incidence of Noncardia Gastric Cancer in US Adults. *JAMA J. Am. Med. Assoc.* **2010**, *303*, 1723–1728. [CrossRef]
18. Bergquist, J.R.; Leiting, J.L.; Habermann, E.B.; Cleary, S.P.; Kendrick, M.L.; Smoot, R.L.; Nagorney, D.M.; Truty, M.J.; Grotz, T.E. Early-Onset Gastric Cancer Is a Distinct Disease with Worrisome Trends and Oncogenic Features. *Surgery* **2019**, *166*, 547–555. [CrossRef]
19. Medina-Franco, H.; Heslin, M.J.; Cortes-Gonzalez, R. Clinicopathological Characteristics of Gastric Carcinoma in Young and Elderly Patients: A Comparative Study. *Ann. Surg. Oncol.* **2000**, *7*, 515–519. [CrossRef]
20. Schildberg, C.W.; Abba, M.; Merkel, S.; Agaimy, A.; Dimmler, A.; Schlabrakowski, A.; Croner, R.; Leupold, J.H.; Hohenberger, W.; Allgayer, H. Gastric Cancer Patients Less than 50 Years of Age Exhibit Significant Downregulation of E-Cadherin and CDX2 Compared to Older Reference Populations. *Adv. Med. Sci.* **2014**, *59*, 142–146. [CrossRef] [PubMed]
21. Pérez-Carbonell, L.; Ruiz-Ponte, C.; Guarinos, C.; Alenda, C.; Payá, A.; Brea, A.; Egoavil, C.M.; Castillejo, A.; Barberá, V.M.; Bessa, X.; et al. Comparison between Universal Molecular Screening for Lynch Syndrome and Revised Bethesda Guidelines in a Large Population-Based Cohort of Patients with Colorectal Cancer. *Gut* **2012**, *61*, 865–872. [CrossRef] [PubMed]
22. Kim, H.S.; Kim, J.W.; Hwang, I.G.; Lee, H.S.; Kim, W.H. Expression of DNA Damage Response Markers in Early-Onset or Familial Gastric Cancers. *Asian Pac. J. Cancer Prev.* **2019**, *20*, 1369–1376. [CrossRef]

23. Giryes, A.; Oweira, H.; Mannhart, M.; Decker, M.; Abdel-Rahman, O. Exploring the Differences between Early-Onset Gastric Cancer and Traditional-Onset Gastric Cancer. *J. Gastrointest. Oncol.* **2018**, *9*, 1157–1163. [CrossRef] [PubMed]
24. Machlowska, J.; Baj, J.; Sitarz, M.; Maciejewski, R.; Sitarz, R. Gastric Cancer: Epidemiology, Risk Factors, Classification, Genomic Characteristics and Treatment Strategies. *Int. J. Mol. Sci.* **2020**, *21*, 4012. [CrossRef] [PubMed]
25. Merchant, S.J.; Kim, J.; Choi, A.H.; Sun, V.; Chao, J.; Nelson, R. A Rising Trend in the Incidence of Advanced Gastric Cancer in Young Hispanic Men. *Gastric Cancer* **2017**, *20*, 226–234. [CrossRef]
26. Li, S.; Rexin, P.; Qin, Z.; Changbo, C.; Guanghui, C.; Luyao, W.; Voelker, H.U.; Stauch, G. Different Incidence of Early-Onset Gastric Carcinoma Depending on Ethnicity: Preliminary Results of a Hospital in Liangshan. *Sci. World J.* **2020**, 1–5. [CrossRef]
27. Adar, T.; Friedman, M.; Rodgers, L.H.; Shannon, K.M.; Zukerberg, L.R.; Chung, D.C. Gastric Cancer in Lynch Syndrome Is Associated with Underlying Immune Gastritis. *J. Med. Genet.* **2019**, *56*, 844–845. [CrossRef]
28. Kim, J.; Braun, D.; Ukaegbu, C.; Dhingra, T.G.; Kastrinos, F.; Parmigiani, G.; Syngal, S.; Yurgelun, M.B. Clinical Factors Associated With Gastric Cancer in Individuals With Lynch Syndrome. *Clin. Gastroenterol. Hepatol.* **2020**, *18*, 830–837.e1. [CrossRef]
29. Zhang, Q.; Wang, L.; Ni, S.; Tan, C.; Cai, X.; Huang, D.; Sheng, W. Clinicopathological Features and Prognostic Value of Mismatch Repair Protein Deficiency in Gastric Cancer. *Int. J. Clin. Exp. Pathol.* **2018**, *11*, 2579–2587.
30. Hayden, J.D.; Cawkwell, L.; Sue-Ling, H.; Johnston, D.; Dixon, M.F.; Quirke, P.; Martin, I.G. Assessment of Microsatellite Alterations in Young Patients with Gastric Adenocarcinoma. *Cancer* **1997**, *79*, 684–687. [CrossRef]
31. Bass, A.J.; Thorsson, V.; Shmulevich, I.; Reynolds, S.M.; Miller, M.; Bernard, B.; Hinoue, T.; Laird, P.W.; Curtis, C.; Shen, H.; et al. Comprehensive Molecular Characterization of Gastric Adenocarcinoma. *Nature* **2014**, *513*, 202–209.
32. Moreira, L.; Muñoz, J.; Cuatrecasas, M.; Quintanilla, I.; Leoz, M.L.; Carballal, S.; Ocaña, T.; Lõpez-Cerõn, M.; Pellise, M.; Castellví-Bel, S.; et al. Prevalence of Somatic Mutl Homolog 1 Promoter Hypermethylation in Lynch Syndrome Colorectal Cancer. *Cancer* **2015**, *121*, 1395–1404. [CrossRef] [PubMed]
33. Tedaldi, G.; Pirini, F.; Tebaldi, M.; Zampiga, V.; Cangini, I.; Danesi, R.; Arcangeli, V.; Ravegnani, M.; Khouzam, R.A.; Molinari, C.; et al. Multigene Panel Testing Increases the Number of Loci Associated with Gastric Cancer Predisposition. *Cancers* **2019**, *11*, 1340. [CrossRef] [PubMed]
34. Aronson, M.; Swallow, C.; Govindarajan, A.; Semotiuk, K.; Cohen, Z.; Kaurah, P.; Velsher, L.; Ambus, I.; Buckley, K.; Forster-Gibson, C.; et al. Germline Variants and Phenotypic Spectrum in a Canadian Cohort of Individuals with Diffuse Gastric Cancer. *Curr. Oncol.* **2020**, *27*, e182–e190. [CrossRef] [PubMed]
35. Vogelaar, I.P.; Van Der Post, R.S.; Van Krieken, J.H.J.M.; Spruijt, L.; Van Zelst-Stams, W.A.G.; Kets, C.M.; Lubinski, J.; Jakubowska, A.; Teodorczyk, U.; Aalfs, C.M.; et al. Unraveling Genetic Predisposition to Familial or Early Onset Gastric Cancer Using Germline Whole-Exome Sequencing. *Eur. J. Hum. Genet.* **2017**, *25*, 1246–1252. [CrossRef] [PubMed]
36. Vos, E.L.; Salo-Mullen, E.E.; Tang, L.H.; Schattner, M.; Yoon, S.S.; Gerdes, H.; Markowitz, A.J.; Mandelker, D.; Janjigian, Y.; Offitt, K.; et al. Indications for Total Gastrectomy in CDH1 Mutation Carriers and Outcomes of Risk-Reducing Minimally Invasive and Open Gastrectomies. *JAMA Surg.* **2020**, *155*, 1050–1057. [CrossRef]
37. Gamble, L.A.; Heller, T.; Davis, J.L. Hereditary Diffuse Gastric Cancer Syndrome and the Role of CDH1: A Review. *JAMA Surg.* **2021**, *156*, 387–392. [CrossRef]
38. Milne, A.N.; Sitarz, R.; Carvalho, R.; Carneiro, F.; Offerhaus, G.J.A. Early Onset Gastric Cancer: On the Road to Unraveling Gastric Carcinogenesis. *Curr. Mol. Med.* **2007**, *7*, 15–28. [CrossRef]
39. Rugge, M.; Busatto, G.; Cassaro, M.; Shiao, Y.H.; Russo, V.; Leandro, G.; Avellini, C.; Fabiano, A.; Sidoni, A.; Covacci, A. Patients Younger than 40 Years with Gastric Carcinoma: Helicobacter Pylori Genotype and Associated Gastritis Phenotype. *Cancer* **1999**, *85*, 2506–2511. [CrossRef]

cancers

Review

Pediatric Neuroendocrine Neoplasms: Rare Malignancies with Incredible Variability

Jennifer T. Castle [1], Brittany E. Levy [2] and Aman Chauhan [3,*]

1 Department of Surgery, Markey Cancer Center, University of Kentucky, 800 Rose Street, Lexington, KY 40536, USA
2 Department of Surgery, University of Kentucky, 800 Rose Street, Lexington, KY 40536, USA
3 Department of Internal Medicine-Medical Oncology, Markey Cancer Center, University of Kentucky, 800 Rose Street, Lexington, KY 40536, USA
* Correspondence: amanchauhan@uky.edu or dramanchauhan@gmail.com

Simple Summary: Neuroendocrine neoplasms are increasing in incidence at a remarkable rate meaning more providers are encountering them in both adult and pediatric patients. This classification of neoplasm encompasses a wide range of different malignancies with a variety of symptoms at presentation and each treated differently. Additionally, over the past few years there has been a change in classification of these neoplasms and a variety of changes and advances in how they are treated. Given this and their rarity in pediatric patients, healthcare providers may not be familiar with these changes. Our goal with this review was to provide an overview of all the most commonly encountered forms of neuroendocrine neoplasms in pediatric patients with up to date recommendations so any healthcare provider can quickly and accurately acclimate themselves.

Abstract: Neuroendocrine neoplasms (NENs) encompass a variety of neuroendocrine tumors (NETs) and neuroendocrine carcinomas (NECs) which can arise anywhere in the body. While relatively rare in the pediatric population, the incidence of NENs has increased in the past few decades. These neoplasms can be devastating if not diagnosed and treated early, however, symptoms are variable and can be indolent for many years. There is a reported median of 10 years from the appearance of the first symptoms to time of diagnosis. Considering some of these neoplasms have a mortality rate as high as 90%, it is crucial healthcare providers are aware of NENs and remain vigilant. With better provider education and easily accessible resources for information about these neoplasms, awareness can be improved leading to earlier disease recognition and diagnosis. This manuscript aims to provide an overview of both the most common NENs as well as the rarer NENs with high lethality in the pediatric population. This review provides up to date evidence and recommendations, encompassing recent changes in classification and advances in treatment modalities, including recently completed and ongoing clinical trials.

Keywords: pediatric; neuroendocrine neoplasms; neuroendocrine tumors; neuroendocrine carcinomas

Citation: Castle, J.T.; Levy, B.E.; Chauhan, A. Pediatric Neuroendocrine Neoplasms: Rare Malignancies with Incredible Variability. *Cancers* **2022**, *14*, 5049. https://doi.org/10.3390/cancers14205049

Academic Editor: David Wong

Received: 21 September 2022
Accepted: 13 October 2022
Published: 15 October 2022

1. Introduction

Neuroendocrine neoplasms (NENs) originate from neuroendocrine cells which can be found throughout the body. As such, NENs can develop anywhere neuroendocrine cells are in the body, but are most commonly found in the lungs, pancreas, and gastrointestinal tract [1]. The classification of NENs is inconsistent amongst organ system of origination. The World Health Organization (WHO) has provided some clarity and more consistent guidelines for the grading of NENs in their most updated reports and has been reflected in the recent National Comprehensive Cancer Network (NCCN) guideline updates [2]. However, classification and grading for NENs remain organ specific to a certain degree. Gastroenteropancreatic NENs are divided into two groups based on differentiation: neuroendocrine tumors (NETs) and neuroendocrine carcinomas (NECs). NECs are poorly differentiated tumors with mitotic rate greater than 20 mitoses per 2 mm^2 and Ki-67 index greater than 20%. NETs are well-differentiated lesions which encompass the remainder of the NENs and are further divided into three grades: grade 1 (no necrosis, mitotic rate less than 2 per 2 mm^2, Ki-67 less than 3%), grade 2 (necrosis present or mitotic rate of 2–20 mitosis per 2 mm^2, Ki-67 between 3–20%), and grade 3 (well differentiated with mitotic rate of greater than 20 mitosis per 2 mm^2 or Ki-67 greater than 20%) [3,4]. Many of the other organ systems follow this same classification system. Contrarily, bronchopulmonary NENs, while still separated based on mitotic rate and differentiation, are named differently and irrespective of Ki-67 proliferation index (typical, atypical, and carcinoma) (Table 1) [3,5,6]. In 2018, the International Agency for Research on Cancer and a WHO expert consensus meeting proposed a uniform classification scheme irrespective of site of origin to create more consistency across organ systems [7]. While their proposed system to dichotomize all NENs into either NET of NEC for every organ system has not been implemented, this may represent the future direction with subsequent classification updates.

Separate but related to classification, the staging systems for some of the organ systems have also changed over the past few years. The 8th edition of the American Joint Committee on Cancer (AJCC) separated pancreatic NEN staging to reflect the difference in tumor biology and prognosis between pancreatic NETs and pancreatic NEC. In this revision, pancreatic NEC still fall under the staging system of other exocrine pancreatic tumors, however, a new staging system was proposed for pancreatic NETs in accordance with their more benign nature [8]. As the scientific community is better able to delineate the clinical course and prognoses of NENs, their classification and staging systems continue to evolve.

Given the complexity and rarity within this disease in the pediatric population, healthcare providers may not be familiar with NENs. Our goal with this review was to provide an overview of commonly encountered forms of NENs in pediatric patients with up to date recommendations so any healthcare provider can quickly and accurately acclimate themselves.

Table 1. Classification of neuroendocrine neoplasms adapted from the WHO classifications up through 2022 as described previously [5,6].

	Grade	Terminology/Differentiation/ Location	Mitotic Rate (per 2 mm^2)	Ki-67 (%)
Neuroendocrine Neoplasms	**Low**	Grade 1, well-differentiated NET (extra-thoracic)	<2	<3
		Typical Carcinoid (Thoracic)	<2	-
	Intermediate	Grade 2, well-differentiated NET (extra-thoracic)	2–20	3–20
		Atypical Carcinoid (Thoracic)	2–10	-
	High	Grade 2, well-differentiated NET	>20	>20
		Poorly differentiated Carcinoma (small cell and large cell)	>20 (thoracic > 10)	>20

2. Epidemiology

NETs are more thoroughly characterized in the adult population with pediatric focused research lacking due to its rarity in this population. Overall, the incidence of NENs has grown by almost 7-fold from 1973 to 2012 [9]. Given the advances and more frequent use of imaging, the incidence of pediatric NENs is suspected to have continued to increase since 2011, similar to the adult population, but an updated pediatric-specific epidemiological study is lacking in the literature. Based on the National Cancer Institute's Surveillance, Epidemiology, and End Results (SEER) data, pediatric NENs are most-commonly found in the gastrointestinal tract (i.e., appendix), lung (one of the most common primary lung neoplasms in children) and breast, but have been reported in a variety of locations throughout the body (Figure 1) [10–12]. However, the appendix has historically been thought of as the overwhelmingly most common NEN in children, which has been demonstrated to account for 80% of all pediatric NENs in other published studies [13].

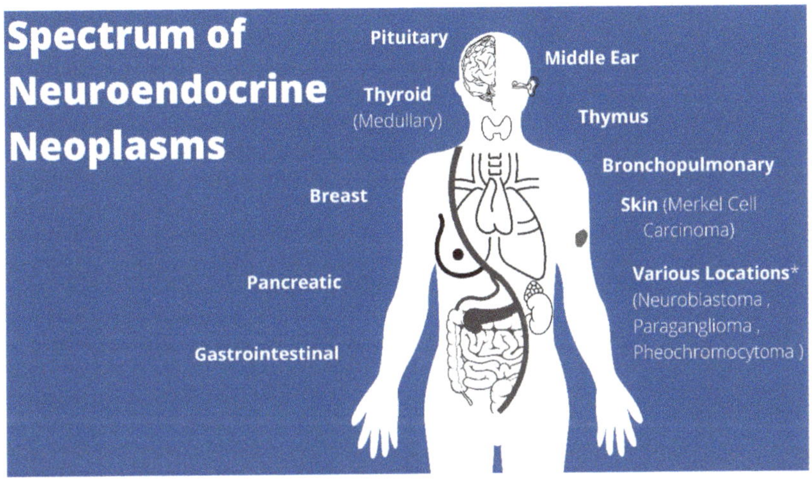

Figure 1. Illustration of some of the common and less common neuroendocrine neoplasms encountered in the pediatric and adult populations (* most commonly originate from the adrenal glands; paragangliomas are extra-adrenal).

3. Diagnosis

3.1. Presentation

NENs are notoriously indolent with vague initial symptoms that present a median of almost ten years prior to diagnosis [14]. This is ubiquitous in both the adult and pediatric population leading to delays in diagnosis in both. It is important to recognize that patients diagnosed as young adults may have had symptoms for many years, in which case they may have developed symptoms in adolescence and childhood. Given their proclivity to go unnoticed for years, a substantial proportion of patients have metastatic disease at presentation. Overall in children, 22% have regional spread, 10% have distant disease, and 5% have an unknown primary site at time of diagnosis [12]. Similar to the spectrum of diseases that are encompassed in pediatric NENs, metastatic rate ranges greatly depending on the location of the primary lesion. Appendiceal NENs rarely ever metastasize where as other gastroenteropancreatic NENs are metastatic at time of diagnosis in 50% of children [15].

Symptoms of NENs, when they do manifest, are dependent on the location as well as the functional secretory status of the tumor. Appendiceal NENs in children are typically non-secretory and diagnosed on pathology after resection for presumed appendicitis. These children typically present rather acutely with only days of symptoms consistent with appendicitis (e.g., abdominal pain, fever, and anorexia) [16–18]. Additionally, imaging, whether it be ultrasound or CT, is usually consistent with acute appendicitis [19]. Similarly, they can have an associated leukocytosis, elevated c-reactive protein, and/or thrombocytosis. But, around half of the reported cases had no laboratory abnormalities [19]. Along the natural course of treatment for presumed appendicitis, they undergo appendectomy and pathology later reveals appendiceal NEN. Other extra-appendiceal gastrointestinal NENs, can present with nondescript abdominal pain and/or diarrhea [15]. Bronchopulmonary NENs are notorious for presenting as a cough with recurrent pneumonias that then progress to wheezing and difficulty breathing [20–22]. These symptoms can go on incorrectly diagnosed as asthma given the similarity in symptoms and age of onset in the pediatric population [11].

When functional, NENs have symptoms secondary to hormone hypersecretion. The classically associated hypersecretion syndrome associated with NENs is carcinoid syndrome. Symptoms of carcinoid syndrome include diarrhea, difficulty breathing, and flushing typically secondary to hypersecretion of serotonin [23]. Previously it was thought that carcinoid syndrome only evolved from hepatic metastasis allowing for direct systemic vascular injection of the hormones from the tumors [23]. However, carcinoid syndrome and one of its more clinically detrimental complications, carcinoid heart disease, has been reported in patients with no evidence of hepatic metastasis [24,25]. Although well documented in the adult population, there are few cases of carcinoid syndrome in the pediatric literature [13,26,27]. In addition to carcinoid syndrome, some NENs are associated with ectopic Cushing's syndrome, a condition characterized by excess glucocorticoids from hypersecretion of adrenocorticotropic hormone (ACTH). Symptoms include growth deceleration, truncal obesity, facial plethora, high blood pressure, and weight gain [28]. Despite NENs being the most common cause of ectopic ACTH overproduction in the pediatric population, occurrences remain infrequent [29]. Of the pediatric reported cases, most ectopic ACTH-producing NENs are bronchial or pancreatic [22,30]. Pituitary adenomas are the overall most-common source of ACTH overproduction (non-ectopic) in children and have recently been reclassified as neuroendocrine neoplasms by the WHO [28,31,32].

The pancreas is the site of a multitude of different functional NENs. Gastrin production results in Zollinger-Ellison syndrome which can present as severe, recurrent epigastric pain and malabsorption from gastric and duodenal ulcers and diarrhea. NENs in the duodenum and the pancreas have been documented to produce gastrin [15,33]. In children, the gastrin-producing NENs (gastrinoma) and insulin-producing NENs (insulinoma) are the most common functional pancreatic NENs [15]. Similar to adults, insulinomas in children present with hypoglycemic symptoms (i.e., sweating, light-headedness, confusion, palpitations) [34,35]. Other functional pancreatic NENs have been described in the literature but are exceedingly rare. VIPomas, defined by hypersecretion of vasoactive intestinal peptide (VIP), present with severe dehydration and malabsorption from uncontrollable diarrhea, but are much rarer in the pediatric population [36]. Somatostatinomas (hypersecretion of somatostatin) may present in the pancreas or the duodenum but are rare even in the adult population. There has only been sparsely documented occurrences in the pediatric and young adult population [37,38]. Presenting symptoms of somatostatinoma include abdominal pain, gastrointestinal bleeding, and jaundice [37]. With similar infrequency, glucagonomas (pancreatic NEN with hypersecretion of glucagon) have very few documented cases in the literature in pediatric and young adult patients [39–41]. Necrolytic migratory erythema, a classic skin rash associated with glucagonomas, is often a presenting feature as well as diabetes and weight loss [42].

Pheochromocytomas and paragangliomas are catecholamine-secreting NENs and responsible for less than 5% of pediatric patients with hypertension [43]. Up to 90% of pediatric patients with a pheochromocytoma or paraganglioma present with hypertension [43]. Some children also experience headaches, palpitations, tremors, and flushing [44]. Pheochromocytomas and paragangliomas originate from the same cell type but differ in that pheochromocytomas are of the adrenal gland and paragangliomas are extra-adrenal tumors. Neuroblastomas, another neuroendocrine tumor frequently originating from the adrenal gland but can also be found anywhere in the sympathetic nervous system, is also associated with catecholamine secretion [45]. However, patients with neuroblastoma typically present with an asymptomatic mass or symptoms secondary to the location of the mass rather than symptoms of excess catecholamines [46]. This can include constipation from bowel compression, hypertension from renal artery compression, and scoliosis with or without neurosensory and motor symptoms from spinal cord compression [47–50]. Neuroblastomas are heterogenous with courses ranging from benign to very aggressive. Often discussed with neuroblastomas are ganglioneuromas and ganglioneuroblastomas, which are benign and mixed-form tumors, respectively. These tumors are cellularly similar to neuroblastomas but are beyond the scope of this review [51].

Quite distinct from the previously mentioned NENs are Merkel Cell Carcinomas of the skin. This is a very aggressive skin cancer with high mortality rate [52]. There are documented cases in the pediatric population demonstrating variable presentations from simple skin dysplasia to subcuticular masses [53]. There is evidence to suggest a more aggressive nature in children as they are 3 times more likely to present with metastatic disease compared to adults [54]. However, due to the extreme rarity in this population, large scale studies to further examine these findings are lacking. This disease remains an individually case reportable event.

3.2. Familial Syndromes

Some children are at a greater risk of developing certain neuroendocrine neoplasms compared to the general pediatric population based on inherited genetic mutations. This includes but is not limited to multiple endocrine neoplasia (MEN) 1, MEN2A, MEN2B, familial medullary thyroid carcinoma (FMTC), Von-Hippel Lindau (vHL) syndrome, Neurofibromatosis type 1 (NF-1), and hereditary paraganglioma-pheochromocytoma syndrome. MEN1 results from a genetic mutation in the MEN1 gene which encodes the Menin protein [55]. Around 15% of patients with MEN1 are diagnosed in childhood and should be screened for the associated malignancies. Children with MEN1 most frequently present with pituitary (prolactin-secreting) and pancreatic (gastrinomas, insulinomas, and non-functional) NENs, in addition to primary hyperparathyroidism which is not currently classified as an NEN [56]. Children with MEN1 are not limited to NENs in these locations as there have also been documented cases of MEN1 patients with thymic, bronchopulmonary and adrenal NENs [56,57]. Given the heterogeneity in disease manifestation in this genetic syndrome, clinical management is complex and patient specific [58].

Akin to MEN1, MEN2A and 2B are also inherited in an autosomal dominant pattern, except they encompass the diseases that arise from a mutation in the RET gene [59]. There is essentially a complete penetrance of medullary thyroid cancer in MEN2A and MEN2B patients which may present as a thyroid nodule [59]. As such, children with MEN2A or 2B are evaluated for a medullary thyroid cancer risk profile and may undergo prophylactic thyroidectomy at or before 5 years old [60]. MEN2A and MEN2B patients are also at increased risk for pheochromocytomas and should begin screening at 11–16 years old depending on risk profile, determined by the specific RET mutation (Table 2) [2,60].

vHL results from a mutation in the VHL tumor suppressor gene which encodes the VHL protein, a regulator of angiogenesis [61]. Among many other malignancies, children with vHL are at an increased risk for pheochromocytomas, paragangliomas, and pancreatic NENs [15,62,63]. Children with NF-1 (mutation of the NF1 gene) are also at an increased risk for pheochromocytomas [43]. Pancreatic NENs are rare in NF-1 patients but have been reported in young adults with NF-1 [37,64].

Table 2. Screening recommendations in children with known familial syndromes at increased risk for developing different NENs. Modified from the 2015 American Thyroid Association guidelines and the 2021 NCCN Neuroendocrine and Adrenal Tumors guidelines [2,60]. (CT = computed tomography, MRI = magnetic resonance imaging, EUS = endoscopic ultrasound, CEA = carcinoembryonic antigen).

<table>
<tr><th></th><th>Syndrome</th><th>NEN</th><th>NEN Percent Occurrence</th><th colspan="2">Screening Recommendations</th></tr>
<tr><td rowspan="9">Familial Syndromes</td><td rowspan="2">MEN 1</td><td>Pancreatic (gastrinoma, insulinoma)</td><td>20–80%</td><td colspan="2">– chromogranin-A, pancreatic polypeptide, glucagon, VIP annually starting at 8 years old
– fasting gastrin annually starting at 20 years old
– consider abdominal CT, MRI, or EUS every 3–5 years starting at 20 years old</td></tr>
<tr><td>Pituitary Adenoma (prolactinoma)</td><td>30–40%</td><td colspan="2">– serum prolactin, IGF-1, fasting glucose and insulin annually starting at 5 years old
– head MRI every 3–5 years starting at 5 years old</td></tr>
<tr><td rowspan="5">MEN 2A/2B</td><td rowspan="3">Medullary Thyroid Cancer</td><td rowspan="3">≥98%</td><td>Highest Risk</td><td>Prophylactic thyroidectomy at or before 1 year old with physical exam, neck ultrasound, serum calcitonin/CEA every 6 months for 1 year and annually thereafter (*serum calcitonin confounding in infants as normally elevated)</td></tr>
<tr><td>High Risk</td><td>Physical exam, neck ultrasound, serum calcitonin annually starting at 3 years old; Prophylactic thyroidectomy at or before 5 years old based on serum calcitonin followed by physical exam, neck ultrasound, serum calcitonin, and CEA every 6 months for 1 year and annually thereafter</td></tr>
<tr><td>Moderate Risk</td><td>Physical exam, serum calcitonin every 6 months for 1 year and annually thereafter if calcitonin remains normal; prophylactic thyroidectomy when calcitonin levels elevated</td></tr>
<tr><td rowspan="2">Pheochromocytoma</td><td rowspan="2">≥50%</td><td>Highest/ High Risk</td><td>Free plasma metanephrines/normetanephrines or 24 h urine fractionated metanephrines annually starting at 11 years old. Adrenal imaging with CT/MRI if elevated</td></tr>
<tr><td>Moderate Risk</td><td>Free plasma metanephrines/normetanephrines or 24 h urine fractionated metanephrines annually starting at 16 years old. Adrenal imaging with CT/MRI if elevated</td></tr>
<tr><td rowspan="3">Von-Hippel Lindau</td><td>Pheochromocytoma</td><td>10–20%</td><td colspan="2" rowspan="3">– Blood pressure at all medical visits starting at 2 years old
– Free plasma metanephrines/normetanephrines or 24 h urine fractionated metanephrines annually starting at 5 years old
– abdominal MRI or CT with and without IV contrast every 2 years starting at 15 years old</td></tr>
<tr><td>Paraganglioma</td><td>10–20%</td></tr>
<tr><td>Pancreatic</td><td>5–17%</td></tr>
</table>

3.3. Biochemical Work Up

Overall, NENs encompass a wide array of diseases and can manifest anytime in childhood and adulthood with a diverse symptom profile. Clinicians should be thoughtful in evaluating children with abnormally persistent symptoms, no matter how vague, as potential for an NEN. A clinical suspicion for an NEN warrants a biochemical diagnostic work-up and, if indicated, imaging. Many NENs are not diagnosed until histologically proven on a pathology sample. There are different markers that distinguish NENs from other types of tumors. Immunohistochemical staining for chromogranin-A (Cg-A) and synaptophysin are used for diagnosis of NEN, but some NENs express different granins than Cg-A potentiating misdiagnosis [3]. Cg-A can also be detected in plasma and serum but some authors have noted issue with low sensitivity (67–93%) and specificity (85%–96%) depending on the diagnostic method employed [65]. Similar issues with low sensitivity but higher specificity has been noted in using neuron-specific enolase for NEN diagnosis [66]. Considering the pitfalls with the currently utilized biochemical tests, there is need for better and more reliable biomarkers. At time of this publication there is an ongoing prospective multi-center trial investigating human circulating progastrin (hPG80) as a biomarker for the monitoring of NENs (NTC04750954). hPG80 has been demonstrated to be elevated in the plasma of patients with a variety of low- and high-grade NENs when compared to healthy controls [67]. However, its wide-spread use is pending validation.

There are a variety of different peptides that can be detected in the blood or urine of patients who present with functional tumors. Urinary 5-Hydroxyindolacetic Acid (5-HIAA) is a metabolite of serotonin which is helpful in the diagnosis of serotonin-producing tumors with higher sensitivity for midgut NENs compared to others [66]. 5-HIAA has high specificity, but serum serotonin and Cg-A can be used in conjunction for diagnosis [68]. Recently, a post-hoc analysis in the CLARINET study showed that 5-HIAA may have use in the diagnosis and monitoring of patients with nonfunctional NENs [69]. However, this has not been adopted into practice guidelines. For medullary thyroid cancer, serum calcitonin measurements are useful in disease detection as well as recurrence screening [60]. The functional pancreatic NENs can also be diagnosed with biochemical work up. Fasting gastrin levels and secretin stimulation tests can be used in the diagnosis of gastrinomas [70]. Classically, it has been advised to stop a proton-pump inhibitor prior to the stimulation test as it confounds the results. However, recent literature suggests this may be unnecessary [71]. Interestingly, the use of proton-pump inhibitors has also been shown to falsely elevate Cg-A levels, confounding that diagnostic test as well [72]. Over 95% of insulinomas can be diagnosed by serial plasma glucose and insulin levels during a 72 h fast showing hyperinsulinemic hypoglycemia [73]. Similarly, somatostatin can be measured to aid in the diagnosis of somatostatinoma and VIP can be measured in the plasma to diagnose a VIPoma in conjunction with hypokalemia and achlorhydria [74,75]. Functional pituitary NENs can be detected by elevated levels of growth hormone, ACTH, IGF1, or prolactin [31].

3.4. Imaging

There are several imaging modalities useful for the diagnosis of NENs. With the increased use of cross-sectional imaging, more and more NENs are incidentally found on computed tomography (CT) and magnetic resonance imaging (MRI) [76]. While this trend has been noted as the etiology of the increased incidence of NENs in the adult population, the same argument does not hold true in the pediatric population where frequent imaging is less likely to occur in accordance with limiting exposure to radiation (i.e., CT). However, when an NEN is suspected in a child or adolescent, undergoing the appropriate diagnostic imaging is important in establishing a timely diagnosis.

Imaging modalities, used alone or in combination with each other, include CT, MRI, positron emission tomography (PET), meta-iodobenzylguanidine (MIBG) scintigraphy, and somatostatin receptor (SSR) scintigraphy. For pituitary NENs, MRI is preferred to distinguish a pituitary lesion from surrounding soft-tissue lesions, whereas CT and the functional imaging studies are of less widespread use but may be appropriate for a select subset of patients with pituitary NENs [77]. In general, CT and MRI alone can help distinguish between well and poorly differentiated NENs based on enhancement, there remains other, more sensitive imaging modalities for the detection and diagnosis of NENs. MIBG is over 90% sensitive for the diagnosis of neuroblastomas, pheochromocytomas, and paragangliomas, but less sensitive in the detection of other NENs (e.g., pancreatic, midgut, non-catecholamine producing tumors) [78,79]. SSR scintigraphy uses a radiolabeled somatostatin analogue (111In-octreotide) to detect NENS by exploiting the elevated expression of somatostatin receptors on the majority of NEN tumor cells. However, SSR-PET/CT and SSR-PET/MRI have improved diagnostic quality compared to scintigraphy alone [80]. In pediatrics, SSR-PET is the imaging modality of choice given its decreased radiation dose and faster study time without compromising diagnostic sensitivity [81]. 68-Gallium-DOTATATE/DOTATOC/DOTANOC are the most commonly used SSR analogues used for diagnostic purposes today [79]. 64-Copper-DOTATATE is another radiolabeled SSR analogue FDA approved for use in the diagnosis of NENs which has been shown to be better at detecting NENs when compared to gallium [82]. In general, SSR-PET is limited in its ability to detect NENs that do not express high levels of somatostatin receptors as well as its ability to distinguish small NENs from surrounding tissue with elevated somatostatin receptors (i.e., pituitary NENs, inflammatory reactions) [77,83]. The benefit of functional imaging with MIBG or SSR-PET is the tumors detectable with these imaging modalities may benefit from therapeutic targeted radioisotope treatments (discussed further in the treatment section below). Additionally, some of these imaging modalities have the ability to distinguish NENs based on differentiation. Well-differentiated NENs demonstrate greater uptake with 68-Gallium-DOTA-peptide PET/CT, whereas poorly-differentiated NENs have low uptake. The reverse is observed with use of 18F-fluorodeoxyglucose (FDG)-PET/CT, an imaging modality which measures glucose metabolism. This phenomenon is well described in gastroenteropancreatic NENs, and has shown to be beneficial in assisting in classifying lesions in tandem with mitotic rate and Ki-67 [84]. In general, the diagnosis of a NEN in a pediatric patient typically involves a combination of these imaging modalities directed by clinical and biochemical work up [2].

3.5. Biopsy

Biopsy can be useful in the diagnosis and characterization of certain NENs. Biopsy can be used in bronchopulmonary NENs if anatomically accessible but may not be sufficient to distinguish between typical and atypical tumors [85,86]. When anatomically accessible, endoscopic ultrasound with fine needle aspiration can be used in gastric, proximal small bowel, and pancreatic NENs [87]. Biopsy does not have reliable sensitivity in pancreatic lesions less than 2 cm and also may mischaracterize neoplasms as Ki-67 indices can vary throughout the tumor itself [88,89]. Biopsy is helpful in the diagnosis of hepatic metastasis and may be necessary if the primary lesion is unknown [90]. Contrarily, biopsy of small bowel NENs are not typically feasible and are not performed but instead are diagnosed through the biopsy of hepatic metastases or surgical excision of the primary lesion [91]. Biopsy of suspected pheochromocytomas and paragangliomas are unique for NENs as it is actively not recommended to biopsy these lesions for diagnosis unless absolutely necessary. This is due to their increased risk of bleeding from hypervascularity, tumor seeding, and hypertensive crisis. If biopsy is necessary, it should only be done so once the patient is satisfactorily α-blocked to reduce the risk of hypertensive crisis [92].

As with any procedure, risk and benefit must be weighed prior to proceeding. If the biochemical and imaging work up is equivocal, biopsy can be beneficial in establishing a diagnosis. However, there are reports of inducing carcinoid crisis and even death from performing biopsies of NENs, but this is very rare [93–95]. Thus, biopsy may not be advisable if the diagnosis is already established and the results from which will not affect treatment planning. In the spirit of personalized medicine, the emergence of a "liquid biopsy" is gaining traction in the evaluation of malignancies. Liquid biopsies have the advantage over the typical biopsy techniques as they are less invasive, faster, and can collect more than sufficient volume of specimen for multiple analyses [96]. The NETest analyzes blood samples for circulating neuroendocrine genomic analytes (i.e., mRNA) and has been used in the evaluation of pheochromocytomas, paragangliomas, bronchopulmonary, gastrointestinal, and pancreatic NENs [97–100]. In the prospective study, NETest had a 99% accuracy in diagnosing NENs [100]. Some have also shown its benefit in predicting response to therapy and assessing for residual disease [101,102]. The role of this study in the pediatric population is uncertain.

4. How to Treat

4.1. Surgery

While complete surgical resection is the ideal treatment for most NENs, careful observation is appropriate for certain subsets of NENs. Infants less than 6 months old with small neuroblastomas can safely be observed as these tumors in this age group can spontaneously regress [103]. While not associated with regression, asymptomatic, non-functional pancreatic NENs less than 2 cm in size can also be observed [104,105]. Some have argued there is an unacceptable risk of disease progression and metastasis with observation [106]. As children have a longer amount of time to potentially develop progression of disease, it may be beneficial to excise these small tumors, but this remains controversial.

Technique for surgical excision is dependent on the location and grade/stage of the disease. Appendiceal NENs are often diagnosed on pathology after appendectomy for presumed appendicitis in children. With guidelines extrapolated from adult observations, it has classically been advised that appendiceal NENs greater than 2 cm should undergo right hemicolectomy due to risk of metastasis with special consideration for radical surgery in other tumors with high-risk features (e.g., high mitotic rate, high Ki-67 proliferation index, incomplete resection, lymph node involvement, and tumor at the base of the appendix) [107,108]. Yet, there are multiple studies in pediatric patients showing no survival advantage nor difference in disease progression for children undergoing simple appendectomy versus right hemicolectomy [109–113]. Thus, making the argument that radical surgery may not be beneficial in this population.

There is an emphasis on early detection for all NENs as complete surgical resection is curative in early locoregional disease [2,114]. This includes partial pneumonectomy (e.g., sleeve resections and bronchoplasty) for bronchopulmonary NENs and partial pancreatectomy (e.g., pancreaticoduodenectomy and distal pancreatectomy) for pancreatic NENs [115]. Even in local disease, surgery for an NEN can be complicated by severe reactions that can occur during surgical resection and anesthesia. Carcinoid crisis causes drastic and sudden hemodynamic instability from the sudden release of vasoactive peptides [116]. The risk of carcinoid crisis during procedures and surgical resection is not negligible as it has a reported incidence of 19%, with greater likelihood in patients with hepatic metastases [116]. While rapid administration of intravenous octreotide is the treatment for carcinoid crisis, recent studies have shown the prophylactic use of octreotide does not decrease the risk of developing perioperative/periprocedural carcinoid crisis [116–118]. Similar to carcinoid crisis, hypertensive crisis can occur in the perioperative/periprocedural time in patients with pheochromocytomas and paragangliomas due to sudden release of catecholamines [119]. Hypertensive crisis has been reported in children [120,121]. As such, children with catecholamine secreting tumors should undergo α-blockade with phenoxybenzamine or doxazosin followed by β-blockade prior to any procedures/operations [122]. Similar to adults, children should be sufficiently α-blocked prior to administration of β-blockers to prevent unopposed α-stimulation [123].

There are a variety of treatment options for metastatic NEN, but unfortunately none are curative like complete surgical resection. In patients with resectable primary lesion and metastatic disease (i.e., hepatic metastases), complete surgical resection of both with cytoreductive surgery is recommended [124]. In the setting of unresectable metastatic disease with resectable primary tumor, debulking by excision of the primary tumor is associated with increased 5-year survival in midgut NEN adult patients [125]. Hepatic metastases can also be managed with transarterial embolization (bland, radiation, or chemotherapeutic), selective internal radiation, or ablation [124,126]. Radiation and radiopharmaceuticals may be effective for primary and metastatic disease depending on the tumor characteristics detailed in the next section. Overall, given the diversity of disease processes encompassed under the classification of NEN, all pediatric patients with NENs should be discussed with a multi-disciplinary team with expertise in the management of pediatric NENs prior to surgical/procedural intervention.

4.2. Radiation Therapy

Radiation therapy and therapeutic radiopharmaceuticals are typically reserved for the treatment of unresectable or residual NENs, or patients not medically appropriate for surgical intervention [127,128]. External beam radiotherapy and/or stereotactic radiosurgery are used in aggressive, residual, and unresectable pituitary NENs, medullary thyroid carcinoma, middle ear NENs, thymic NENs, bronchopulmonary NENs, and gastroenteropancreatic NENs [60,129–134]. It is rare to use radiation therapy as a single therapeutic agent for any NEN but instead used in multimodal treatment plans.

Patients with norepinephrine and somatostatin receptor avid tumors detected with the functional imaging discussed above can be treated with similarly structured therapeutic radiopharmaceuticals. Nonresectable neuroblastomas, pheochromocytomas, and paragangliomas with elevated MIBG uptake are often treated with 131I-MIBG alone or in combination with chemotherapeutics [135–137]. A therapy still in evolution that offers improved outcomes over 131I-MIBG, is peptide receptor radionuclide therapy (PRRT). PRRT uses radiolabeled somatostatin analogues to deliver radiation therapy directly to the tumor. It is suitable for the treatment of NENs which overexpress somatostatin receptors [138]. Two different radiopeptides, 90Y-DOTATOC and 177Lu-DOTATATE, are in use today. PRRT has shown increased overall survival, progression free survival, event free survival, and response to treatment when compared to 131I-MIBG in the treatment of advanced pheochromocytomas and paragangliomas [139]. PRRT has also been used in MIBG refractory neuroblastoma [140–142]. The NETTER-1 trial showed that 177Lu-DOTATATE improved both progression free survival and quality of life [143]. At this time, 177Lu-DOTATATE is not FDA approved for use in children and is only used in clinical trials on a case-by-case basis. Prior to the use of either 131I-MIBG or PRRT, functional imaging proving the tumor's avidity for these radiotherapeutics should be confirmed. Additionally, the NETest may be of use in predicting PRRT response and monitoring disease status during treatment [101,144].

As it stands, these therapies remain a palliative option for the treatment of advanced NENs. It is important to recognize these therapies do have potential adverse effects. Similar to operations and procedures, PRRT can induce carcinoid crisis and MIBG therapy can induce hypertensive crisis in NEN patients [145–148]. Although, these occurrences are documented in adults and it is unclear what the true risk is for pediatric patients.

4.3. Medical Management

Similarly, to how somatostatin analogues are beneficial in imaging and radiotherapy, they can also be used in treatment and symptom control. Octreotide and lanreotide are somatostatin analogues which have long been used to control carcinoid symptoms and reduce disease progression in patients with NENs [149–151]. The CLARINET trial demonstrated significant improvement in progression free-survival of 65.1% at 24 months in NEN patients treated with lanreotide versus only 33.0% in the placebo group [151]. The PROMID trial found octreotide was beneficial in prolonging time to tumor progression [152,153]. There are multiple studies evaluating these agents and others for disease control and symptom management, including the TELESTAR study which found telotristat ethyl improved carcinoid syndrome diarrhea and reduced urinary 5-HIAA levels in patients not well controlled with somatostatin analogues [154]. However, response is dependent on somatostatin receptor status of the tumor and PRRT has been shown to be more effective than somatostatin analogues alone in the treatment of adult NEN patients [143].

Many of the treatment options available to pediatric NEN patients is extrapolated from what works in adults [155]. Various chemotherapeutics (cyclophosphamide, vincristine, dacarbazine, temozolomide, capecitabine, etoposide, cisplatin/carboplatin, everolimus, and mTOR inhibitors) have been used with varying success to treat advanced NENs [60,156–162]. The use of chemotherapeutics in pediatric patients must be decided on an individual basis with consideration of inclusion in clinical trials suited to the location and molecular profile of the tumor.

4.4. Clinical Trials

Due to the rarity of NENs in the pediatric population and the wide variety of organ systems they can originate from, there are few clinical trials that have been conducted or are being conducted specifically in this population beyond those focused on radiopharmaceuticals. In general, pediatric patients with NENs have and continue to be treated in solid tumor clinical trials that involve both pediatric and adult populations typically geared toward other solid malignancies. In a form of concordance with the movement towards personalized medicine, many of these patients are included in clinical trials if they have refractory or relapsed disease with certain genetic mutation profiles that can be targeted with the investigational chemotherapeutics (e.g., MDM2, MDMX, RET, BCL-2, and many others) [163]. Current ongoing therapeutic clinical trials focused on NENs are investigating the clinical efficacy and safety of 177Lu-DOTATATE in children with gastroenteropancreatic NENs, paragangliomas, and neuroblastomas (NCT04711135, NCT03966651) [164–166]. Prior Phase I and II clinical trials with 90Y-DOTATOC and 177Lu-DOTATATE in children with solid tumors have shown minimal dose-limiting toxicities with a good safety profile indicating their safety in this population [167,168]. As these ongoing clinical trial progress to completion, the treatment paradigm of pediatric NENs are expected to change accordingly. In the interim, a pediatric patient with refractory, recurrent, and/or retained neuroendocrine disease should be considered for inclusion in a clinical trial.

5. Long-Term Outcomes

Overall, pediatric cancer patients are living longer with more cancer survivors living well into adulthood [169]. Although many childhood cancers survival rates have seen large improvements over the past few decades, only a very modest improvement in survival has been seen in some pediatric NEN patients. For instance, from 1975 to 2006, 5-year survival for neuroblastomas increased nearly 30 percent whereas over the same period, 5-year survival only increased by 1 percent for other pediatric NEN patients. Overall, pediatric NEN patients do have a better survival compared to adults, but survival varies widely based on site of the primary lesion [6]. Survival for appendiceal NEN is observed to be 100% whereas foregut NEN survival is only 26%, and even worse for those of unknown primary (10.5% observed survival rate) [12]. Given this significant heterogeneity in survival outcomes and the individual rarity for each tumor location, it is not surprising that clinical trials and studies do not capture pediatric NENs as an individual entity like they do for other pediatric malignancies (i.e., leukemias, lymphomas, etc.).

Similar to the wide margin of survival for all pediatric NENs, rate of recurrence is equally tumor and organ specific. Of the multitude of studies on pediatric appendiceal NENs, most report no observed recurrences in their patient populations [109]. One study out of Poland has documented a recurrence after surgical resection (which was subsequently surgically removed, with no further evidence of recurrence on follow-up) [170]. In contrast, other extra-appendiceal gastrointestinal NETs, although rarer than appendiceal NETs in children, have a greater risk of recurrence [13]. Within the adult and pediatric population, recurrence of middle ear NEN is quoted to be 22% [171]. There are multiple small case series and retrospective reviews on bronchial NENs in mixed pediatric and adult populations where no recurrences were detected and others finding a recurrence rate of 2–27% depending on histological subtype (atypical 7.9 times more likely to recur than typical) [13,22,172]. The recurrence rate for medullary thyroid cancer even after total thyroidectomy is 9–12% in children but varies greatly in timing of thyroidectomy and genetic predisposition [60,173]. Children with MEN 2 mutations who undergo prophylactic thyroidectomy at younger ages have lower risk of recurrent or persistent disease [173]. Hence the recommendation for prophylactic thyroidectomy at or before 5 years old in children with high risk MEN2 mutations [60]. Recurrence of NENs can occur even 50 years after initial diagnosis making life-long surveillance a necessity for childhood NEN survivors [174].

The North American and British Childhood Cancer Survivor Studies (CCSS and BCCSS) have been instrumental in defining the long-term outcomes for a number of pediatric malignancies, but unfortunately NENs have not been characterized through this study [175,176]. Thus, the long-term outcomes of survivors of pediatric NENs as they live into adulthood are relatively unknown. Through the childhood cancer survivor studies, it was found that childhood cancer survivors develop more chronic health conditions (cardiac, musculoskeletal, neurologic, endocrine, and gastrointestinal) and are at higher risk of developing a subsequent malignancy compared to the general population [169,177]. Childhood cancer survivors also face social and economic disadvantages in life as they are less likely to graduate college and less likely to have full-time employment [178,179]. Almost 15% of childhood cancer survivors develop posttraumatic stress symptoms that can impede quality of life [180]. Many of these risks differ based on the primary malignancy, however the treatment modalities employed during childhood also confer risk of developing these outcomes in adulthood. Although these studies do not pertain to pediatric NEN patients in particular, it is unlikely pediatric NEN patients are exempt from these trends as the treatment modalities are similar to other childhood malignancies. These patients should be monitored appropriately for outcomes of the like well into adulthood.

6. Challenges and Opportunities

Like most rare diseases, relatively low incidence has been a significant deterrent to advancements in clinical drug development in NENs. Many clinical trials are often terminated early due to lack of enrollment. NEN translational research also suffers from lack of easily accessible high quality pre-clinical models. Last but not the least, dichotomization of NENs among various site-specific disease groups has led to a lack of common terminology, classification, and management framework. While we acknowledge the above mentioned challenges, we are also optimistic about the road ahead. The several fold increase in the incidence of NENs has garnered attention of not only the pharmaceutical industry but also the National Cancer Institute (NCI) and has resulted in a significant upsurge in interventional therapeutic clinical trials [181]. Consensus is being generated to homogenize terminology and classification of various subsets of NENs. Our understanding of the molecular underpinning of NENs has substantially improved in the last decade and molecular characterization of NENs is not only being considered to classify the morphologic subtypes of NENs but will also pave the way for relevant precision medicine clinical trials in the future.

7. Conclusions

NENs can present in a variety of ways with outcomes which range from benign to very aggressive. It is crucial healthcare providers of all levels in all specialties be aware of how NENs can present given the potential for high morbidity in delayed diagnoses. As our understanding of this disease continues to progress, the management of NENs continues to evolve. This review provides an overview of all the commonly encountered forms of NENs in pediatric patients with up to date recommendations so any healthcare provider can quickly and accurately acclimate themselves.

Author Contributions: J.T.C. contributed to conceptualization, resource curation, writing-original draft preparation, review and editing. B.E.L. contributed to writing-original draft preparation, review and editing. A.C. contributed to conceptualization, writing-review and editing. All authors have read and agreed to the published version of the manuscript.

Funding: This research was supported by the NIH Training Grant T32CA160003.

Conflicts of Interest: The authors declare no conflict of interest.

References

1. Man, D.; Wu, J.; Shen, Z.; Zhu, X. Prognosis of patients with neuroendocrine tumor: A SEER database analysis. *Cancer Manag. Res.* **2018**, *10*, 5629–5638. [CrossRef] [PubMed]

2. Shah, M.H.; Goldner, W.S.; Benson, A.B.; Bergsland, E.; Blaszkowsky, L.S.; Brock, P.; Chan, J.; Das, S.; Dickson, P.V.; Fanta, P.; et al. Neuroendocrine and Adrenal Tumors, Version 2.2021, NCCN Clinical Practice Guidelines in Oncology. *J. Natl. Compr. Canc. Netw.* **2021**, *19*, 839–868. [CrossRef]
3. Bellizzi, A.M. Immunohistochemistry in the diagnosis and classification of neuroendocrine neoplasms: What can brown do for you? *Hum. Pathol.* **2020**, *96*, 8–33. [CrossRef]
4. Carvão, J.; Dinis-Ribeiro, M.; Pimentel-Nunes, P.; Libânio, D. Neuroendocrine Tumors of the Gastrointestinal Tract: A Focused Review and Practical Approach for Gastroenterologists. *GE Port J. Gastroenterol.* **2021**, *28*, 336–348. [CrossRef] [PubMed]
5. Corbett, V.; Arnold, S.; Anthony, L.; Chauhan, A. Management of Large Cell Neuroendocrine Carcinoma. *Front. Oncol.* **2021**, *11*, 653162. [CrossRef] [PubMed]
6. Rindi, G.; Mete, O.; Uccella, S.; Basturk, O.; La Rosa, S.; Brosens, L.A.A.; Ezzat, S.; de Herder, W.W.; Klimstra, D.S.; Papotti, M.; et al. Overview of the 2022 WHO Classification of Neuroendocrine Neoplasms. *Endocr. Pathol.* **2022**, *33*, 115–154. [CrossRef] [PubMed]
7. Rindi, G.; Klimistra, D.S.; Abedi-Ardekani, B.; Asa, S.L.; Bosman, F.T.; Brambilla, E.; Busam, K.J.; de Krijger, R.R.; Dietel, M.; El-Naggar, A.K.; et al. A common classification framework for neuroendocrine neoplasms: An International Agency for Research on Cancer (IARC) and World Health Organization (WHO) expert consensus proposal. *Mod. Pathol.* **2018**, *31*, 1770–1786. [CrossRef]
8. Amin, M.B.; Edge, S.B.; Greene, F.L.; Byrd, D.R.; Brookland, R.K.; Washington, M.K.; Gershenwald, J.E.; Comptom, C.C.; Hess, K.R.; Sullivan, D.C.; et al. *AJCC Cancer Staging Manual*, 8th ed.; Springer: Cham, Switzerland; New York, NY, USA, 2017.
9. Dasari, A.; Shen, C.; Halperin, D.; Zhao, B.; Zhou, S.; Xu, Y.; Shih, T.; Yao, J.C. Trends in the Incidence, Prevalence, and Survival Outcomes in Patients with Neuroendocrine Tumors in the United States. *JAMA Oncol.* **2017**, *3*, 1335–1342. [CrossRef] [PubMed]
10. Giuseppucci, C.; Reusmann, A.; Giubergia, V.; Barrias, C.; Kruger, A.; Siminovich, M.; Botto, H.; Cadario, M.; Boglione, M.; Strambach, J.; et al. Primary lung tumors in children: 24 years of experience at a referral center. *Pediatr. Surg. Int.* **2016**, *32*, 451–457. [CrossRef] [PubMed]
11. Dishop, M.K.; Kuruvilla, S. Primary and metastatic lung tumors in the pediatric population: A review and 25-year experience at a large children's hospital. *Arch. Pathol. Lab. Med.* **2008**, *132*, 1079–1103. [CrossRef]
12. Navalkele, P.; O'Dorisio, M.S.; O'Dorisio, T.M.; Zamba, G.K.; Lynch, C.F. Incidence, survival, and prevalence of neuroendocrine tumors versus neuroblastoma in children and young adults: Nine standard SEER registries, 1975–2006. *Pediatr. Blood Cancer* **2011**, *56*, 50–57. [CrossRef] [PubMed]
13. Degnan, A.J.; Tocchio, S.; Kurtom, W.; Tadros, S.S. Pediatric neuroendocrine carcinoid tumors: Management, pathology, and imaging findings in a pediatric referral center. *Pediatr. Blood Cancer.* **2017**, *64*, e26477. [CrossRef] [PubMed]
14. Vinik, A.I.; Chaya, C. Clinical Presentation and Diagnosis of Neuroendocrine Tumors. *Hematol. Oncol. Clin. North Am.* **2016**, *30*, 21–48. [CrossRef]
15. Khanna, G.; O'Dorisio, S.M.; Menda, Y.; Kirby, P.; Kao, S.; Sato, Y. Gastroenteropancreatic neuroendocrine tumors in children and young adults. *Pediatr. Radiol.* **2008**, *38*, 251–259. [CrossRef]
16. Kim, S.S.; Kays, D.W.; Larson, S.D.; Islam, S. Appendiceal carcinoids in children–management and outcomes. *J. Surg. Res.* **2014**, *192*, 250–253. [CrossRef]
17. Amr, B.; Froghi, F.; Edmond, M.; Haq, K.; Thengungal-Kochupapy, R. Management and outcomes of appendicular neuroendocrine tumours: Retrospective review with 5-year follow-up. *Eur. J. Surg. Oncol.* **2015**, *41*, 1243–1246. [CrossRef] [PubMed]
18. Ranaweera, C.; Brar, A.; Somers, G.R.; Sheikh, F.; Pierro, A.; Zani, A. Management of pediatric appendiceal carcinoid: A single institution experience from 5000 appendectomies. *Pediatr. Surg. Int.* **2019**, *35*, 1427–1430. [CrossRef] [PubMed]
19. Lobeck, I.N.; Jeste, N.; Geller, J.; Pressey, J.; von Allmen, D. Surgical management and surveillance of pediatric appendiceal carcinoid tumor. *J. Pediatr. Surg.* **2017**, *52*, 925–927. [CrossRef] [PubMed]
20. Rojas, Y.; Shi, Y.X.; Zhang, W.; Beierle, E.A.; Doski, J.J.; Goldfarb, M.; Goldin, A.B.; Gow, K.W.; Langer, M.; Vasudevan, S.A.; et al. Primary malignant pulmonary tumors in children: A review of the national cancer data base. *J. Pediatr. Surg.* **2015**, *50*, 1004–1008. [CrossRef] [PubMed]
21. Madafferi, S.; Catania, V.D.; Accinni, A.; Boldrini, R.; Inserra, A. Endobronchial tumor in children: Unusual finding in recurrent pneumonia, report of three cases. *World J. Clin. Pediatr.* **2015**, *4*, 30–34. [CrossRef] [PubMed]
22. Potter, S.L.; HaDuong, J.; Okcu, F.; Wu, H.; Chintagumpala, M.; Venkatramani, R. Pediatric Bronchial Carcinoid Tumors: A Case Series and Review of the Literature. *J. Pediatr. Hematol. Oncol.* **2019**, *41*, 67–70. [CrossRef] [PubMed]
23. Ito, T.; Lee, L.; Jensen, R.T. Carcinoid-syndrome: Recent advances, current status and controversies. *Curr. Opin. Endocrinol. Diabetes Obes.* **2018**, *25*, 22–35. [CrossRef] [PubMed]
24. Ram, P.; Penalver, J.L.; Lo, K.B.U.; Rangaswami, J.; Pressman, G.S. Carcinoid Heart Disease: Review of Current Knowledge. *Tex. Heart Inst. J.* **2019**, *46*, 21–27. [CrossRef] [PubMed]
25. Bhattacharyya, S.; Toumpanakis, C.; Caplin, M.E.; Davar, J. Analysis of 150 patients with carcinoid syndrome seen in a single year at one institution in the first decade of the twenty-first century. *Am. J. Cardiol.* **2008**, *101*, 378–381. [CrossRef] [PubMed]
26. Cogen, J.D.; Swanson, J.; Ong, T. Endobronchial Carcinoid and Concurrent Carcinoid Syndrome in an Adolescent Female. *Case Rep. Pediatr.* **2016**, *2016*, 2074970. [CrossRef]
27. Spunt, S.L.; Pratt, C.B.; Rao, B.N.; Pritchard, M.; Jenkins, J.J.; Hill, D.A.; Cain, A.M.; Pappo, A.S. Childhood carcinoid tumors: The St Jude Children's Research Hospital experience. *J Pediatr. Surg.* **2000**, *35*, 1282–1286. [CrossRef]

28. Lodish, M.B.; Keil, M.F.; Stratakis, C.A. Cushing's Syndrome in Pediatrics: An Update. *Endocrinol. Metab. Clin. North Am.* **2018**, *47*, 451–462. [CrossRef]
29. More, J.; Young, J.; Reznik, Y.; Raverot, G.; Borson-Chazot, F.; Rohmer, V.; Baudin, E.; Coutant, R.; Tabarin, A.; Groupe Français des Tumeurs Endocrines (GTE). Ectopic ACTH syndrome in children and adolescents. *J. Clin. Endocrinol. Metab.* **2011**, *96*, 1213–1222. [CrossRef]
30. Saxena, R.; Pathak, M.; Shukla, R.; Sinha, A.; Elhence, P.; Bharti, J.N.; Khera, P. Bronchial Carcinoid Tumour as a Rare Cause of Cushing's Syndrome in Children: A Case Report and Review of Literature. *J. Clin. Res. Pediatr. Endocrinol.* **2020**, *12*, 340–346. [CrossRef]
31. Trouillas, J.; Jaffrain-Rea, M.L.; Vasiljevic, A.; Raverot, G.; Roncaroli, F.; Villa, C. How to Classify the Pituitary Neuroendocrine Tumors (PitNET)s in 2020. *Cancers* **2020**, *12*, 514. [CrossRef]
32. Asa, S.L.; Mete, O.; Perry, A.; Osamura, R.Y. Overview of the 2022 WHO Classification of Pituitary Tumors. *Endocr. Pathol.* **2022**, *33*, 6–26. [CrossRef]
33. Cholewa, D.; Waldschmidt, J.; Hoffmann, K.; Bäder, M.; Zimmer, T.; Scherübl, H.; Riecken, E.O.; Wiedenmann, B. A 7-year-old child with primary tumour localisation in the distal duodenum–new imaging procedures for an improved diagnosis. *Eur. J. Pediatr.* **1997**, *156*, 568–571. [CrossRef] [PubMed]
34. Liang, M.; Jiang, J.; Dai, H.; Hong, X.; Han, X.; Cong, L.; Tong, A.; Li, F.; Luo, Y.; Liu, W.; et al. Robotic enucleation for pediatric insulinoma with MEN1 syndrome: A case report and literature review. *BMC Surg.* **2018**, *18*, 44. [CrossRef] [PubMed]
35. Padidela, R.; Fiest, M.; Arya, V.; Smith, V.V.; Ashworth, M.; Rampling, D.; Newbould, M.; Batra, G.; James, J.; Wright, N.B.; et al. Insulinoma in childhood: Clinical, radiological, molecular and histological aspects of nine patients. *Eur. J. Endocrinol.* **2014**, *170*, 741–747. [CrossRef]
36. Bonilla Gonzalez, C.; Rusinque, J.; Uribe, C.; Carias, A.; Contreras, M.L. Pancreatic VIPoma as a Differential Diagnosis in Chronic Pediatric Diarrhea: A Case Report and Review of the Literature. *J. Med. Cases.* **2021**, *12*, 195–201. [CrossRef] [PubMed]
37. Garbrecht, N.; Anlauf, M.; Schmitt, A.; Henopp, T.; Sipos, B.; Raffel, A.; Eisenberger, C.F.; Knoefel, W.T.; Pavel, M.; Fottner, C.; et al. Somatostatin-producing neuroendocrine tumors of the duodenum and pancreas: Incidence, types, biological behavior, association with inherited syndromes, and functional activity. *Endocr. Relat. Cancer.* **2008**, *15*, 229–241. [CrossRef] [PubMed]
38. Thannberger, P.; Wilhelm, J.M.; Derragui, A.; Saraceni, O.; Kieffer, P. Maladie de Von Recklinghausen associée à un somatostatinome pancréatique [Von Recklinghausen's disease associated with pancreatic somatostatinoma]. *Presse Med.* **2001**, *30*, 1741–1743.
39. Kindmark, H.; Sundin, A.; Granberg, D.; Dunder, K.; Skogseid, B.; Janson, E.T.; Welin, S.; Oberg, K.; Eriksson, B. Endocrine pancreatic tumors with glucagon hypersecretion: A retrospective study of 23 cases during 20 years. *Med. Oncol.* **2007**, *24*, 330–337. [CrossRef]
40. Luber, A.J.; Ackerman, L.S.; Culpepper, K.S.; Buschmann, C.M.; Koep, L.J. Paediatric necrolytic migratory erythema as a presenting sign of glucagonoma syndrome. *Br. J. Dermatol.* **2016**, *174*, 1092–1095. [CrossRef]
41. Wermers, R.A.; Fatourechi, V.; Wynne, A.G.; Kvols, L.K.; Lloyd, R.V. The glucagonoma syndrome. Clinical and pathologic features in 21 patients. *Medicine* **1996**, *75*, 53–63. [CrossRef]
42. Ito, T.; Igarashi, H.; Jensen, R.T. Pancreatic neuroendocrine tumors: Clinical features, diagnosis and medical treatment: Advances. *Best Pract. Res. Clin. Gastroenterol.* **2012**, *26*, 737–753. [CrossRef] [PubMed]
43. Bholah, R.; Bunchman, T.E. Review of Pediatric Pheochromocytoma and Paraganglioma. *Front Pediatr.* **2017**, *5*, 155. [CrossRef] [PubMed]
44. Peard, L.; Cost, N.G.; Saltzman, A.F. Pediatric pheochromocytoma: Current status of diagnostic imaging and treatment procedures. *Curr. Opin. Urol.* **2019**, *29*, 493–499. [CrossRef]
45. Verly, I.R.N.; van Kuilenburg, A.B.P.; Abeling, N.G.G.M.; Goorden, S.M.I.; Fiocco, M.; Vaz, F.M.; van Noesel, M.M.; Zwaan, C.M.; Kaspers, G.L.; Merks, J.H.M.; et al. Catecholamines profiles at diagnosis: Increased diagnostic sensitivity and correlation with biological and clinical features in neuroblastoma patients. *Eur. J. Cancer.* **2017**, *72*, 235–243. [CrossRef]
46. Swift, C.C.; Eklund, M.J.; Kraveka, J.M.; Alazraki, A.L. Updates in Diagnosis, Management, and Treatment of Neuroblastoma. *Radiographics* **2018**, *38*, 566–580. [CrossRef] [PubMed]
47. Benya, E. A 13-month-old girl with constipation. *Pelvic Neuroblastoma. Pediatr. Ann.* **2005**, *34*, 680–682. [CrossRef] [PubMed]
48. Marchal, A.L.; Hoeffel, J.C.; Freyd, S.; Schmitt, M.; Olive, D.; Fays, J. Hypertension arterielle par compression extrinseque d'origine tumorale de l'artere renale chez l'enfant [Arterial hypertension caused by extrinsic compression of the renal artery of tumor origin in a child]. *Pediatrie* **1986**, *41*, 475–480.
49. De Martino, L.; Spennato, P.; Vetrella, S.; Capasso, M.; Porfito, C.; Ruotolo, S.; Abate, M.E.; Cinalli, G.; Quaglietta, L. Symptomatic malignant spinal cord compression in children: A single-center experience. *Ital. J. Pediatr.* **2019**, *45*, 80. [CrossRef]
50. Nakazawa, M.; Crook, S.; Horan, J. Suprarenal Neuroblastoma Presenting in a Child With Infantile Scoliosis. *Pediatr. Blood Cancer.* **2016**, *63*, 748–749. [CrossRef] [PubMed]
51. Lonergan, G.J.; Schwab, C.M.; Suarez, E.S.; Carlson, C.L. Neuroblastoma, ganglioneuroblastoma, and ganglioneuroma: Radiologic-pathologic correlation. *Radiographics.* **2002**, *22*, 911–934. [CrossRef] [PubMed]
52. Becker, J.C.; Stang, A.; DeCaprio, J.A.; Cerroni, L.; Lebbé, C.; Veness, M.; Nghiem, P. Merkel cell carcinoma. *Nat. Rev. Dis. Primers.* **2017**, *3*, 17077. [CrossRef] [PubMed]
53. Köksal, Y.; Toy, H.; Talim, B.; Unal, E.; Akçören, Z.; Cengiz, M. Merkel cell carcinoma in a child. *J. Pediatr. Hematol. Oncol.* **2009**, *31*, 359–361. [CrossRef] [PubMed]

54. Paulson, K.G.; Nghiem, P. One in a hundred million: Merkel cell carcinoma in pediatric and young adult patients is rare but more likely to present at advanced stages based on US registry data. *J. Am. Acad. Dermatol.* **2019**, *80*, 1758–1760. [CrossRef] [PubMed]
55. Brandi, M.L.; Agarwal, S.K.; Perrier, N.D.; Lines, K.E.; Valk, G.D.; Thakker, R.V. Multiple Endocrine Neoplasia Type 1: Latest Insights. *Endocr. Rev.* **2021**, *42*, 133–170. [CrossRef] [PubMed]
56. Goudet, P.; Dalac, A.; Le Bras, M.; Cardot-Bauters, C.; Niccoli, P.; Lévy-Bohbot, N.; du Boullay, H.; Bertagna, X.; Ruszniewski, P.; Borson-Chazot, F.; et al. MEN1 disease occurring before 21 years old: A 160-patient cohort study from the Groupe d'etude des Tumeurs Endocrines. *J. Clin. Endocrinol. Metab.* **2015**, *100*, 1568–1577. [CrossRef]
57. Oliveira, A.M.; Tazelaar, H.D.; Wentzlaff, K.A.; Kosugi, N.S.; Hai, N.; Benson, A.; Miller, D.L.; Yang, P. Familial pulmonary carcinoid tumors. *Cancer* **2001**, *91*, 2104–2109. [CrossRef]
58. Pieterman, C.R.; Valk, G.D. Update on the clinical management of multiple endocrine neoplasia type 1. *Clin. Endocrinol.* **2022**, *97*, 409–423. [CrossRef] [PubMed]
59. Wohllk, N.; Schweizer, H.; Erlic, Z.; Schmid, K.W.; Walz, M.K.; Raue, F.; Neumann, H.P.H. Multiple endocrine neoplasia type 2. *Best Pract. Res. Clin. Endocrinol Metab.* **2010**, *24*, 371–387.1. [CrossRef] [PubMed]
60. Wells, S.A., Jr.; Asa, S.L.; Dralle, H.; Elisei, R.; Evans, D.B.; Gagel, R.F.; Lee, N.; Machens, A.; Moley, J.F.; Pacini, F.; et al. Revised American Thyroid Association guidelines for the management of medullary thyroid carcinoma. *Thyroid* **2015**, *25*, 567–610. [CrossRef] [PubMed]
61. Latif, F.; Tory, K.; Gnarra, J.; Yao, M.; Duh, F.M.; Orcutt, M.L.; Stackhouse, T.; Kuzmin, I.; Modi, W.; Geil, L.; et al. Identification of the von Hippel-Lindau disease tumor suppressor gene. *Science* **1993**, *260*, 1317–1320. [CrossRef] [PubMed]
62. Launbjerg, K.; Bache, I.; Galanakis, M.; Bisgaard, M.L.; Binderup, M.L.M. von Hippel-Lindau development in children and adolescents. *Am. J. Med. Genet. A* **2017**, *173*, 2381–2394. [CrossRef]
63. Varshney, N.; Kebede, A.A.; Owusu-Dapaah, H.; Lather, J.; Kaushik, M.; Bhullar, J.S. A Review of Von Hippel-Lindau Syndrome. *J. Kidney Cancer VHL* **2017**, *4*, 20–29. [CrossRef] [PubMed]
64. Shinozaki, H.; Sasakura, Y.; Shinozaki, S.; Terauchi, T.; Matsui, J.; Kobayashi, K.; Lefor, A.K.; Ogata, Y. Cystic Pancreatic Neuroendocrine Tumor in a Patient with Neurofibromatosis Type 1. *Case Rep. Gastroenterol.* **2021**, *15*, 108–114. [CrossRef]
65. Gut, P.; Czarnywojtek, A.; Fischbach, J.; Bączyk, M.; Ziemnicka, K.; Wrotkowska, E.; Gryczyńska, M. Ruchala, M. Chromogranin A—Unspecific neuroendocrine marker. Clinical Utility and potential diagnostic pitfalls. *Arch. Med. Sci.* **2016**, *12*, 1–9. [CrossRef]
66. Eriksson, B.; Oberg, K.; Stridsberg, M. Tumor markers in neuroendocrine tumors. *Digestion* **2000**, *62*, 33–38. [CrossRef] [PubMed]
67. Chauhan, A.; Prieur, A.; Kolesar, J.; Arnold, S.; Payen, L.; Mahi, Y.; Vire, B.; Sands, M.; Evers, B.M.; Joubert, D.; et al. hPG80 (Circulating Progastrin), a Novel Blood-Based Biomarker for Detection of Poorly Differentiated Neuroendocrine Carcinoma and Well Differentiated Neuroendocrine Tumors. *Cancers* **2022**, *14*, 863. [CrossRef] [PubMed]
68. Subash, N.; Papali, M.M.; Bahadur, K.P.; Avanthika, C.; Jhaveri, S.; Thannir, S.; Joshi, M.; Valisekka, S.S. Recent Advances in the Diagnosis and Management of Carcinoid Syndrome. *Dis. Mon.* **2022**, *68*, 10134. [CrossRef] [PubMed]
69. Pavel, M.E.; Phan, A.T.; Wolin, E.M.; Mirakhur, B.; Liyanage, N.; Lowenthal, S.P.; Fisher, G.A., Jr.; Vinik, A.I.; CLARINET Study Investigators. Effect of Lanreotide Depot/Autogel on Urinary 5-Hydroxyindoleacetic Acid and Plasma Chromogranin A Biomarkers in Nonfunctional Metastatic Enteropancreatic Neuroendocrine Tumors. *Oncologist* **2019**, *24*, 463–474. [CrossRef] [PubMed]
70. Berna, M.J.; Hoffmann, K.M.; Serrano, J.; Gibril, F.; Jensen, R.T. Serum gastrin in Zollinger-Ellison syndrome: I. Prospective study of fasting serum gastrin in 309 patients from the National Institutes of Health and comparison with 2229 cases from the literature. *Medicine* **2006**, *85*, 295–330. [CrossRef]
71. Bhattacharya, S.; Blau, J.E.; Cochran, C.; Auh, S.; Weinstein, L.S.; Jensen, R.T.; Wank, S. Validity of Secretin Stimulation Testing on Proton Pump Inhibitor Therapy for Diagnosis of Zollinger-Ellison Syndrome. *Am. J. Gastroenterol.* **2021**, *116*, 2216–2221. [CrossRef] [PubMed]
72. Korse, C.M.; Muller, M.; Taal, B.G. Discontinuation of proton pump inhibitors during assessment of chromogranin A levels in patients with neuroendocrine tumours. *Br. J. Cancer.* **2011**, *105*, 1173–1175. [CrossRef] [PubMed]
73. Donegan, D.; Jakubikova, I.; Vella, A. Anthropometric Features Are Not Predictive of 72-Hour Fast Duration in Insulinomas. *Endocr. Pract.* **2017**, *23*, 923–928. [CrossRef] [PubMed]
74. Kanakis, G.; Kaltsas, G. Biochemical markers for gastroenteropancreatic neuroendocrine tumours (GEP-NETs). *Best Pract. Res. Clin. Gastroenterol.* **2012**, *26*, 791–802. [CrossRef] [PubMed]
75. Una Cidon, E. Vasoactive intestinal peptide secreting tumour: An overview. *World J. Gastrointest. Oncol.* **2022**, *14*, 808–819. [CrossRef]
76. Rosenberg, A.M.; Friedmann, P.; Del Rivero, J.; Libutti, S.K.; Laird, A.M. Resection versus expectant management of small incidentally discovered nonfunctional pancreatic neuroendocrine tumors. *Surgery* **2016**, *159*, 302–309. [CrossRef] [PubMed]
77. Bashari, W.A.; Senanayake, R.; Fernández-Pombo, A.; Gillet, D.; Koulouri, O.; Powlson, A.S.; Matys, T.; Scoffings, D.; Cheow, H.; Mendichovszky, I.; et al. Modern imaging of pituitary adenomas. *Best Pract. Res. Clin. Endocrinol. Metab.* **2019**, *33*, 101278. [CrossRef]
78. Jacobson, A.F.; Deng, H.; Lombard, J.; Lessig, H.J.; Black, R.R. 123I-meta-iodobenzylguanidine scintigraphy for the detection of neuroblastoma and pheochromocytoma: Results of a meta-analysis. *J. Clin. Endocrinol. Metab.* **2010**, *95*, 2596–2606. [CrossRef]
79. Haddad, T.; Fard-Esfahani, A.; Vali, R. A review of pediatric neuroendocrine tumors, their detection, and treatment by radioisotopes. *Nucl. Med. Commun.* **2021**, *42*, 21–31. [CrossRef]

80. Shah, S.; Purandare, N.; Agrawal, A.; Rangarajan, V. A pictoral review on somatostatin receptor scintigraphy in neuroendocrine tumors: The role of multimodality imaging with SRS and GLUT receptor imaging with FDG PET-CT. *Indian J. Radiol. Imaging* **2012**, *22*, 267–275. [CrossRef]
81. Hope, T.A.; Bergsland, E.k.; Bozkurt, M.F.; Graham, M.; Heaney, A.P.; Herrman, K.; Howe, J.R.; Kulke, M.H.; Kunz, P.L.; Mailman, J.; et al. Appropriate Use Criteria for Somatostatin Receptor PET Imaging in Neuroendocrine Tumors. *J. Nucl. Med.* **2018**, *59*, 66–74. [CrossRef]
82. Johnbeck, C.B.; Knigge, U.; Loft, A.; Berthelsen, A.K.; Mortensen, J.; Oturai, P.; Langer, S.W.; Elema, D.R.; Kjaer, A. Head-to-Head Comparison of (64)Cu-DOTATATE and (68)Ga-DOTATOC PET/CT: A Prospective Study of 59 Patients with Neuroendocrine Tumors. *J. Nucl. Med.* **2017**, *58*, 451–457. [CrossRef] [PubMed]
83. Velikyan, I. Prospective of (68)Ga Radionuclide Contribution to the Development of Imaging Agents for Infection and Inflammation. *Contrast Media Mol. Imaging* **2018**, 9713691.
84. Kaewput, C.; Vinjamuri, S. Role of Combined 68Ga DOTA-Peptides and 18F FDG PET/CT in the Evaluation of Gastroenteropancreatic Neuroendocrine Neoplasms. *Diagnostics* **2022**, *12*, 280. [CrossRef]
85. Dixon, R.K.; Britt, E.J.; Netzer, G.A.; Afshar, M.; Burke, A.; Liu, S.; Jean, J.; Shah, N.G. Ten-year Single Center Experience of Pulmonary Carcinoid Tumors and Diagnostic Yield of Bronchoscopic Biopsy. *Lung* **2016**, *194*, 905–910. [CrossRef]
86. Yeh, Y.C.; Chou, T.Y. Pulmonary neuroendocrine tumors: Study of 90 cases focusing on clinicopathological characteristics, immunophenotype, preoperative biopsy, and frozen section diagnoses. *J. Surg. Oncol.* **2014**, *109*, 280–286. [CrossRef] [PubMed]
87. Ahmed, M. Gastrointestinal neuroendocrine tumors in 2020. *World J. Gastrointest. Oncol.* **2020**, *12*, 791–807. [CrossRef] [PubMed]
88. Heidsma, C.M.; Tsilimigras, D.I.; Rocha, F.; Abbott, D.E.; Fields, R.; Smith, P.M.; Poultsides, G.A.; Cho, C.; van Eijck, C.; van Dijkum, E.N.; et al. Clinical relevance of performing endoscopic ultrasound-guided fine-needle biopsy for pancreatic neuroendocrine tumors less than 2 cm. *J. Surg. Oncol.* **2020**, *122*, 1393–1400. [CrossRef]
89. Tatsumoto, S.; Kodama, Y.; Sakurai, Y.; Shinohara, T.; Katanuma, A.; Maguchi, H. Pancreatic neuroendocrine neoplasm: Correlation between computed tomography enhancement patterns and prognostic factors of surgical and endoscopic ultrasound-guided fine-needle aspiration biopsy specimens. *Abdom. Imaging* **2013**, *38*, 358–366. [CrossRef]
90. Frilling, A.; Modlin, I.M.; Kidd, M.; Russell, C.; Breitenstein, S.; Salem, R.; Kwekkeboom, D.; Lau, W.; Klersy, C.; Vilgrain, V.; et al. Recommendations for management of patients with neuroendocrine liver metastases. *Lancet Oncol.* **2014**, *15*, e8–e21. [CrossRef]
91. Howe, J.R.; Cardona, K.; Fraker, D.L.; Kebebew, E.; Untch, B.R.; Wang, Y.Z.; Law, C.H.; Liu, E.H.; Kim, M.K.; Menda, Y.; et al. The Surgical Management of Small Bowel Neuroendocrine Tumors: Consensus Guidelines of the North American Neuroendocrine Tumor Society. *Pancreas* **2017**, *46*, 715–731. [CrossRef]
92. NANETS Guidelines 2021 Edition. North American Neuroendocrine Tumor Society. Available online: https://nanets.net/images/guidelines/2021_NANETS_Guidelines_Compendium.pdf (accessed on 14 July 2022).
93. Bissonnette, R.T.; Gibney, R.G.; Berry, B.R.; Buckley, A.R. Fatal carcinoid crisis after percutaneous fine-needle biopsy of hepatic metastasis: Case report and literature review. *Radiology* **1990**, *174*, 751–752. [CrossRef] [PubMed]
94. Karmy-Jones, R.; Valliéres, E. Carcinoid crisis after biopsy of a bronchial carcinoid. *Ann. Thorac Surg.* **1993**, *56*, 1403–1405. [CrossRef]
95. Jang, S.; Schmitz, J.J.; Atwell, T.D.; Welch, T.L.; Hobday, T.J.; Adamo, D.A.; Moynagh, M.R. Percutaneous Image-Guided Core Needle Biopsy of Neuroendocrine Tumors: How Common Is Intraprocedural Carcinoid Crisis? *J. Vasc. Interv. Radiol.* **2021**, *32*, 745–751. [CrossRef] [PubMed]
96. Poulet, G.; Massias, J.; Taly, V. Liquid Biopsy: General Concepts. *Acta Cytol.* **2019**, *63*, 449–455. [CrossRef]
97. Modlin, I.M.; Kidd, M.; Malczewska, A.; Drozdov, I.; Bodei, L.; Matar, S.; Chung, K.M. The NETest: The Clinical Utility of Multigene Blood Analysis in the Diagnosis and Management of Neuroendocrine Tumors. *Endocrinol. Metab. Clin. North Am.* **2018**, *47*, 485–504. [CrossRef]
98. Pacak, K.; Kidd, M.; Meuter, L.; Modlin, I.M. A novel liquid biopsy (NETest) identifies paragangliomas and pheochromocytomas with high accuracy. *Endocr. Relat. Cancer.* **2021**, *28*, 731–744. [CrossRef] [PubMed]
99. Filosso, P.L.; Oberg, K.; Malczewska, A.; Lewczuk, A.; Roffinella, M.; Aslanian, H.; Bodei, L. Molecular identification of bronchopulmonary neuroendocrine tumours and neuroendocrine genotype in lung neoplasia using the NETest liquid biopsy. *Eur. J. Cardiothorac. Surg.* **2020**, *57*, 1195–1202. [CrossRef]
100. Malczewska, A.; Witkowska, M.; Makulik, K.; Bocian, A.; Walter, A.; Pilch-Kowalczyk, J.; Zajęcki, W.; Bodei, L.; Oberg, K.E.; Kos-Kudla, B. NETest liquid biopsy is diagnostic of small intestine and pancreatic neuroendocrine tumors and correlates with imaging. *Endocr. Connect.* **2019**, *8*, 442–453. [CrossRef]
101. Bodei, L.; Kidd, M.S.; Singh, A.; van der Zwan, W.A.; Severi, S.; Drozdov, I.A.; Malczewska, A.; Baum, R.P.; Kwekkeboom, D.J.; Paganelli, G.; et al. PRRT neuroendocrine tumor response monitored using circulating transcript analysis: The NETest. *Eur. J. Nucl. Med. Mol. Imaging* **2020**, *47*, 895–906. [CrossRef]
102. Partelli, S.; Andreasi, V.; Muffatti, F.; Schiavo Lena, M.; Falconi, M. Circulating Neuroendocrine Gene Transcripts (NETest): A Postoperative Strategy for Early Identification of the Efficacy of Radical Surgery for Pancreatic Neuroendocrine Tumors. *Ann. Surg. Oncol.* **2020**, *27*, 3928–3936. [CrossRef] [PubMed]
103. Nuchtern, J.G.; London, W.B.; Barnewolt, C.E.; Naranjo, A.; McGrady, P.W.; Geiger, J.D.; Diller, L.; Schmidt, M.L.; Maris, J.M.; Cohn, S.L.; et al. A prospective study of expectant observation as primary therapy for neuroblastoma in young infants: A Children's Oncology Group study. *Ann. Surg.* **2012**, *256*, 573–580. [CrossRef] [PubMed]

104. Falconi, M.; Eriksson, B.; Kaltsas, G.; Bartsch, D.K.; Capdevila, J.; Caplin, M.; Kos-Kudla, B.; Kwekkeboom, D.; Rindi, G.; Klöppel, G.; et al. ENETS Consensus Guidelines Update for the Management of Patients with Functional Pancreatic Neuroendocrine Tumors and Non-Functional Pancreatic Neuroendocrine Tumors. *Neuroendocrinology* **2016**, *103*, 153–171. [CrossRef]
105. Sadot, E.; Reidy-Lagunes, D.L.; Tang, L.H.; Do, R.K.G.; Gonen, M.; D'Angelica, M.; DeMatteo, R.P.; Kingham, T.P.; Koerkamp, B.G.; Untch, B.R.; et al. Observation versus Resection for Small Asymptomatic Pancreatic Neuroendocrine Tumors: A Matched Case-Control Study. *Ann. Surg. Oncol.* **2016**, *23*, 1361–1370. [CrossRef]
106. Aziz, H.; Howe, J.R.; Pawlik, T.M. Surgery vs Observation for Patients with Small Pancreatic Neuroendocrine Tumors. *JAMA Surg.* **2021**, *156*, 412–413. [CrossRef] [PubMed]
107. Kelly, K.J. Management of Appendix Cancer. *Clin. Colon. Rectal. Surg.* **2015**, *28*, 247–255. [CrossRef]
108. Daskalakis, K.; Alexandraki, K.; Kassi, E.; Tsoli, M.; Angelousi, A.; Ragkousi, A.; Kaltsas, G. The risk of lymph node metastases and their impact on survival in patients with appendiceal neuroendocrine neoplasms: A systematic review and meta-analysis of adult and paediatric patients. *Endocrine* **2020**, *67*, 20–34. [CrossRef] [PubMed]
109. Henderson, L.; Fehily, C.; Folaranmi, S.; Kelsey, A.; McPartland, J.; Jawaid, W.B.; Craigie, R.; Losty, P.D. Management and outcome of neuroendocrine tumours of the appendix-a two centre UK experience. *J. Pediatr. Surg.* **2014**, *49*, 1513–1517. [CrossRef] [PubMed]
110. Pawa, N.; Clift, A.K.; Osmani, H.; Drymousis, P.; Cichocki, A.; Flora, R.; Goldin, R.; Patsouras, D.; Baird, A.; Malczewska, A.; et al. Surgical Management of Patients with Neuroendocrine Neoplasms of the Appendix: Appendectomy or More. *Neuroendocrinology* **2018**, *106*, 242–251. [CrossRef]
111. de Lambert, G.; Lardy, H.; Martelli, H.; Orbach, D.; Gauthier, F.; Guérin, F. Surgical Management of Neuroendocrine Tumors of the Appendix in Children and Adolescents: A Retrospective French Multicenter Study of 114 Cases. *Pediatr. Blood Cancer* **2016**, *63*, 598–603. [CrossRef] [PubMed]
112. Fallon, S.C.; Hicks, M.J.; Carpenter, J.L.; Vasudevan, S.A.; Nuchtern, J.G.; Cass, D.L. Management of appendiceal carcinoid tumors in children. *J. Surg. Res.* **2015**, *198*, 384–387. [CrossRef] [PubMed]
113. Malkan, A.D.; Wahid, F.N.; Fernandez-Pineda, I.; Sandoval, J.A. Appendiceal carcinoid tumor in children: Implications for less radical surgery? *Clin. Transl. Oncol.* **2015**, *17*, 197–200. [CrossRef]
114. Zhuge, Y.; Cheung, M.C.; Yang, R.; Eldick, D.; Koniaris, L.G.; Sola, J.E. Pediatric intestinal foregut and small bowel solid tumors: A review of 105 cases. *J. Surg. Res.* **2009**, *156*, 95–102. [CrossRef] [PubMed]
115. Garnier, H.; Loo, C.; Czauderna, P.; Vasudevan, S.A. Pediatric Gastrointestinal Stromal Tumors and Neuroendocrine Tumors: Advances in Surgical Management. *Surg. Oncol. Clin. N. Am.* **2021**, *30*, 219–233. [CrossRef] [PubMed]
116. Xu, A.; Suz, P.; Reljic, T.; Are, A.C.; Kumar, A.; Powers, B.; Strosberg, J.; Denbo, J.W.; Fleming, J.B.; Anaya, D.A. Perioperative Carcinoid Crisis: A Systematic Review and Meta-Analysis. *Cancers* **2022**, *14*, 2966. [CrossRef] [PubMed]
117. Seymour, N.; Sawh, S.C. Mega-dose intravenous octreotide for the treatment of carcinoid crisis: A systematic review. *Can. J. Anaesth.* **2013**, *60*, 492–499. [CrossRef] [PubMed]
118. Wonn, S.M.; Ratzlaff, A.N.; Pommier, S.J.; McCully, B.H.; Pommier, R.F. A prospective study of carcinoid crisis with no perioperative octreotide. *Surgery* **2022**, *171*, 88–93. [CrossRef] [PubMed]
119. Mazza, A.; Armigliato, M.; Marzola, M.C.; Schiavon, L.; Montemurro, D.; Vescovo, G.; Zuin, M.; Chondrogiannis, S.; Ravenni, R.; Opocher, G.; et al. Anti-hypertensive treatment in pheochromocytoma and paraganglioma: Current management and therapeutic features. *Endocrine* **2014**, *45*, 469–478. [CrossRef]
120. Reichardt, P.; Apel, T.W.; Domula, M.; Tröbs, R.B.; Krause, I.; Bierbach, U.; Neumann, H.P.H.; Kiess, W. Recurrent polytopic chromaffin paragangliomas in a 9-year-old boy resulting from a novel germline mutation in the von Hippel-Lindau gene. *J. Pediatr. Hematol. Oncol.* **2002**, *24*, 145–148. [CrossRef]
121. Ambarsari, C.G.; Hidayati, E.L.; Tridjaja, B.; Mochtar, C.A.; Wulandari, H.F.; Harahap, A.S.; Grace, A. Silent Hypertensive Crisis in an Adolescent: First Case Report of Pediatric Pheochromocytoma from Indonesia. *Glob. Pediatr. Health* **2021**, *8*, 2333794X211015484. [CrossRef] [PubMed]
122. Raina, R.; Mahajan, Z.; Sharma, A.; Chakraborty, R.; Mahajan, S.; Sethi, S.K.; Kapur, G.; Kaelber, D. Hypertensive Crisis in Pediatric Patients: An Overview. *Front. Pediatr.* **2020**, *8*, 588911. [CrossRef] [PubMed]
123. Havekes, B.; Romijn, J.A.; Eisenhofer, G.; Adams, K.; Pacak, K. Update on pediatric pheochromocytoma. *Pediatr. Nephrol.* **2009**, *24*, 943–950. [CrossRef] [PubMed]
124. Gangi, A.; Howe, J.R. The Landmark Series: Neuroendocrine Tumor Liver Metastases. *Ann. Surg. Oncol.* **2020**, *27*, 3270–3280. [CrossRef] [PubMed]
125. Tsilimigras, D.I.; Ntanasis-Stathopoulos, I.; Kostakis, I.D.; Moris, D.; Schizas, D.; Cloyd, J.M.; Pawlik, T.M. Is Resection of Primary Midgut Neuroendocrine Tumors in Patients with Unresectable Metastatic Liver Disease Justified? A Systematic Review and Meta-Analysis. *J. Gastrointest. Surg.* **2019**, *23*, 1044–1054. [CrossRef] [PubMed]
126. Harring, T.R.; Nguyen, N.T.; Goss, J.A.; O'Mahony, C.A. Treatment of liver metastases in patients with neuroendocrine tumors: A comprehensive review. *Int. J. Hepatol.* **2011**, *2011*, 154541. [CrossRef] [PubMed]
127. Kim, S.J.; Kim, J.W.; Han, S.W.; Oh, D.Y.; Lee, S.H.; Kim, D.W.; Im, S.A.; Kim, T.Y.; Seog Heo, D.; Bang, Y.J. Biological characteristics and treatment outcomes of metastatic or recurrent neuroendocrine tumors: Tumor grade and metastatic site are important for treatment strategy. *BMC Cancer* **2010**, *10*, 448. [CrossRef]

128. Chen, K.S.; Lawhn-Heath, C.; Behr, S.; Juarez, R.; Whitman, J.; Paciorek, A.; Nakakura, E.K.; Fidelman, N.; Feng, M.U.S.; Bergsland, E.K.; et al. Outcomes after high-dose radiation in the management of neuroendocrine neoplasms. *PLoS ONE* **2021**, *16*, e0252574. [CrossRef] [PubMed]
129. Fleseriu, M.; Auchus, R.; Bancos, I.; Ben-Shlomo, A.; Bertherat, J.; Biermasz, N.R.; Boguszewski, C.L.; Bronstein, M.D.; Buchfelder, M.; Carmichael, J.D.; et al. Consensus on diagnosis and management of Cushing's disease: A guideline update. *Lancet Diabetes Endocrinol.* **2021**, *9*, 847–875. [CrossRef]
130. Mehta, G.U.; McCutcheon, I.E. Stereotactic Radiosurgery for Pituitary Carcinoma. *Acta Neurochir. Suppl.* **2021**, *128*, 43–49.
131. van der Lans, R.J.L.; Engel, M.S.D.; Rijken, J.A.; Hensen, E.F.; Bloemena, E.; van der Torn, M.; Leemans, C.R.; Smit, C.F.G.M. Neuroendocrine neoplasms of the middle ear: Unpredictable tumor behavior and tendency for recurrence. *Head Neck.* **2021**, *43*, 1848–1853.
132. Ruffini, E.; Oliaro, A.; Novero, D.; Campisi, P.; Filosso, P.L. Neuroendocrine tumors of the thymus. *Thorac. Surg. Clin.* **2011**, *21*, 13–23. [CrossRef]
133. Thapa, P.; Parghane, R.; Basu, S. (177)Lu-DOTATATE Peptide Receptor Radionuclide Therapy in Metastatic or Advanced and Inoperable Primary Neuroendocrine Tumors of Rare Sites. *World J. Nucl. Med.* **2017**, *16*, 223–228. [PubMed]
134. Chan, D.L.; Thompson, R.; Lam, M.; Pavlakis, N.; Hallet, J.; Law, C.; Singh, S.; Myrehaug, S. External Beam Radiotherapy in the Treatment of Gastroenteropancreatic Neuroendocrine Tumours: A Systematic Review. *Clin. Oncol.* **2018**, *30*, 400–408. [CrossRef] [PubMed]
135. Carrasquillo, J.A.; Chen, C.C.; Jha, A.; Pacak, K.; Pryma, D.A. Lin FI. Systemic Radiopharmaceutical Therapy of Pheochromocytoma and Paraganglioma. *J. Nucl. Med.* **2021**, *62*, 1192–1199. [CrossRef] [PubMed]
136. Fishbein, L.; Del Rivero, J.; Else, T.; Howe, J.R.; Asa, S.L.; Cohen, D.L.; Dahia, P.L.M.; Fraker, D.L.; Goodman, K.A.; Hope, T.A.; et al. The North American Neuroendocrine Tumor Society Consensus Guidelines for Surveillance and Management of Metastatic and/or Unresectable Pheochromocytoma and Paraganglioma. *Pancreas* **2021**, *50*, 469–493. [CrossRef]
137. Ussowicz, M.; Wieczorek, A.; Dluzniewska, A.; Pieczonka, A.; Dębski, R.; Drabko, K.; Goździk, J.; Balwierz, W.; Handkiewicz-Junak, D.; Wachowiak, J. Factors Modifying Outcome After MIBG Therapy in Children with Neuroblastoma-A National Retrospective Study. *Front. Oncol.* **2021**, *11*, 647361. [CrossRef]
138. Camus, B.; Cottereau, A.S.; Palmieri, L.J.; Dermine, S.; Tenenbaum, F.; Brezault, C.; Coriat, R. Indications of Peptide Receptor Radionuclide Therapy (PRRT) in Gastroenteropancreatic and Pulmonary Neuroendocrine Tumors: An Updated Review. *J. Clin. Med.* **2021**, *10*, 1267. [CrossRef] [PubMed]
139. Nastos, K.; Cheung, V.T.F.; Toumpanakis, C.; Navalkissoor, S.; Quigley, A.M.; Caplin, M.; Khoo, B. Peptide Receptor Radionuclide Treatment and (131)I-MIBG in the management of patients with metastatic/progressive phaeochromocytomas and paragangliomas. *J. Surg. Oncol.* **2017**, *115*, 425–434. [CrossRef] [PubMed]
140. Kong, G.; Hofman, M.S.; Murray, W.K.; Wilson, S.; Wood, P.; Downie, P.; Super, L.; Hogg, A.; Eu, P.; Hicks, R.J. Initial Experience with Gallium-68 DOTA-Octreotate PET/CT and Peptide Receptor Radionuclide Therapy for Pediatric Patients with Refractory Metastatic Neuroblastoma. *J. Pediatr. Hematol. Oncol.* **2016**, *38*, 87–96. [CrossRef]
141. Gains, J.E.; Bomanji, J.B.; Fersht, N.L.; Sullivan, T.; D'Souza, D.; Sullivan, K.P.; Aldridge, M.; Waddington, W.; Gaze, M.N. 177Lu-DOTATATE molecular radiotherapy for childhood neuroblastoma. *J. Nucl. Med.* **2011**, *52*, 1041–1047. [CrossRef]
142. Fathpour, G.; Jafari, E.; Hashemi, A.; Dadgar, H.; Shahriari, M.; Zareifar, S.; Jenabzade, A.R.; Vali, R.; Ahmadzadehfar, H.; Assadi, M. Feasibility and Therapeutic Potential of Combined Peptide Receptor Radionuclide Therapy with Intensive Chemotherapy for Pediatric Patients With Relapsed or Refractory Metastatic Neuroblastoma. *Clin. Nucl. Med.* **2021**, *46*, 540–548. [CrossRef]
143. Strosberg, J.R.; Caplin, M.E.; Kunz, P.L.; Ruszniewski, P.B.; Bodei, L.; Hendifar, A.; Mittra, E.; Wolin, E.M.; Yao, J.C.; Pavel, M.E.; et al. (177)Lu-Dotatate plus long-acting octreotide versus highdose long-acting octreotide in patients with midgut neuroendocrine tumours (NETTER-1): Final overall survival and long-term safety results from an open-label, randomised, controlled, phase 3 trial. *Lancet Oncol.* **2021**, *22*, 1752–1763. [CrossRef]
144. Bodei, L.; Kidd, M.; Modlin, I.M.; Severi, S.; Drozdov, I.; Nicolini, S.; Kwekkeboom, D.J.; Krenning, E.P.; Baum, R.P.; Paganelli, G. Measurement of circulating transcripts and gene cluster analysis predicts and defines therapeutic efficacy of peptide receptor radionuclide therapy (PRRT) in neuroendocrine tumors. *Eur. J. Nucl. Med. Mol. Imaging* **2016**, *43*, 839–851. [CrossRef] [PubMed]
145. Dhanani, J.; Pattison, D.A.; Burge, M.; Williams, J.; Riedel, B.; Hicks, R.J.; Reade, M.C. Octreotide for resuscitation of cardiac arrest due to carcinoid crisis precipitated by novel peptide receptor radionuclide therapy (PRRT): A case report. *J. Crit. Care.* **2020**, *60*, 319–322. [CrossRef] [PubMed]
146. Nakano, S.; Tsushima, Y.; Taketomi-Takahashi, A.; Higuchi, T.; Amanuma, M.; Oriuchi, N.; Endo, K. Hypertensive crisis due to contrast-enhanced computed tomography in a patient with malignant pheochromocytoma. *Jpn. J. Radiol.* **2011**, *29*, 449–451. [CrossRef]
147. Siow Ping, L.; Azraai, B.N.; Subashini, R. Peptide receptor radionuclide therapy induced carcinoid crisis: A case report. *Med. J. Malaysia* **2022**, *77*, 128–131. [PubMed]
148. Tapia Rico, G.; Li, M.; Pavlakis, N.; Cehic, G.; Price, T.J. Prevention and management of carcinoid crises in patients with high-risk neuroendocrine tumours undergoing peptide receptor radionuclide therapy (PRRT): Literature review and case series from two Australian tertiary medical institutions. *Cancer Treat. Rev.* **2018**, *66*, 1–6. [CrossRef] [PubMed]

149. Vainas, I.; Drimonitis, A.; Boudina, M.; Kaprara, A.; Iakovou, I.; Salem, N.; Koussis, C. The therapeutic value of SST-A octreotide alone or with adjuvant treatment in patients with advanced medullary thyroid carcinoma and positive (111)In-octreotide scan. *Hell. J. Nucl. Med.* **2005**, *8*, 43–47. [PubMed]
150. Stueven, A.K.; Kayser, A.; Wetz, C.; Amthauer, H.; Wree, A.; Tacke, F.; Wiedenmann, B.; Roderburg, C.; Jann, H. Somatostatin Analogues in the Treatment of Neuroendocrine Tumors: Past, Present and Future. *Int. J. Mol. Sci.* **2019**, *20*, 3049. [CrossRef] [PubMed]
151. Caplin, M.E.; Pavel, M.; Ćwikla, J.B.; Phan, A.T.; Raderer, M.; Sedláčková, E.; Cadiot, G.; Wolin, E.M.; Capdevila, J.; Wall, L.; et al. Lanreotide in metastatic enteropancreatic neuroendocrine tumors. *N. Engl. J. Med.* **2014**, *371*, 224–233. [CrossRef] [PubMed]
152. Rinke, A.; Müller, H.H.; Schade-Brittinger, C.; Klose, K.J.; Barth, P.; Wied, M.; Mayer, C.; Aminossadati, B.; Pape, U.F.; Bläker, M.; et al. Placebo-controlled, double-blind, prospective, randomized study on the effect of octreotide LAR in the control of tumor growth in patients with metastatic neuroendocrine midgut tumors: A report from the PROMID Study Group. *J. Clin. Oncol.* **2009**, *27*, 4656–4663. [CrossRef] [PubMed]
153. Rinke, A.; Wittenberg, M.; Schade-Brittinger, C.; Aminossadati, B.; Ronicke, E.; Gress, T.M.; Müller, H.H.; Arnold, R.; PROMID Study Group. Placebo-Controlled, Double-Blind, Prospective, Randomized Study on the Effect of Octreotide LAR in the Control of Tumor Growth in Patients with Metastatic Neuroendocrine Midgut Tumors (PROMID): Results of Long-Term Survival. *Neuroendocrinology* **2017**, *104*, 26–32. [CrossRef]
154. Kulke, M.H.; Horsch, D.; Caplin, M.E.; Anthony, L.B.; Bergsland, E.; Öberg, K.; Welin, S.; Warner, R.R.P.; Lombard-Bohas, C.; Kunz, P.L.; et al. Telotristat Ethyl, a Tryptophan Hydroxylas Inhibitor for the Treatment of Carcinoid Syndrome. *J. Clin. Oncol.* **2017**, *35*, 14–23. [CrossRef] [PubMed]
155. Dumortier, T.; Heimann, G.; Fink, M. Exposure-response modeling for extrapolation from adult to pediatric patients who differ with respect to prognostic factors: Application to everolimus. *CPT Pharmacomet. Syst. Pharmacol.* **2021**, *10*, 589–598. [CrossRef] [PubMed]
156. Ilanchezhian, M.; Jha, A.; Pacak, K.; Del Rivero, J. Emerging Treatments for Advanced/Metastatic Pheochromocytoma and Paraganglioma. *Curr. Treat Options Oncol.* **2020**, *21*, 85. [CrossRef] [PubMed]
157. Koumarianou, A.; Kaltsas, G.; Kulke, M.H.; Oberg, K.; Strosberg, J.R.; Spada, F.; Galdy, S.; Barberis, M.; Fumagalli, C.; Berruti, A.; et al. Temozolomide in Advanced Neuroendocrine Neoplasms: Pharmacological and Clinical Aspects. *Neuroendocrinology* **2015**, *101*, 274–288. [CrossRef] [PubMed]
158. Peixoto, R.D.; Noonan, K.L.; Pavlovich, P.; Kennecke, H.F.; Lim, H.J. Outcomes of patients treated with capecitabine and temozolamide for advanced pancreatic neuroendocrine tumors (PNETs) and non-PNETs. *J. Gastrointest. Oncol.* **2014**, *5*, 247–252. [CrossRef]
159. Pavel, M.E.; Hainsworth, J.D.; Baudin, E.; Peeters, M.; Hörsch, D.; Winkler, R.E.; Klimovsky, J.; Lebwohl, D.; Jehl, V.; Wolin, E.M.; et al. Everolimus plus octreotide long-acting repeatable for the treatment of advanced neuroendocrine tumours associated with carcinoid syndrome (RADIANT-2): A randomised, placebo-controlled, phase 3 study. *Lancet* **2011**, *378*, 2005–2012. [CrossRef]
160. Yao, J.C.; Shah, M.H.; Ito, T.; Lombard-Bohas, C.; Wolin, E.M.; Cutsem, E.V.; Hobday, T.J.; Okusaka, T.; Capdevila, J.; de Vries, E.G.E.; et al. Everolimus for advanced pancreatic neuroendocrine tumors. *N. Engl. J. Med.* **2011**, *364*, 514–523. [CrossRef] [PubMed]
161. Yao, J.C.; Fazio, N.; Singh, S.; Buzzoni, R.; Carnaghi, C.; Wolin, E.; Tomasek, J.; Raderer, M.; Lahner, H.; Voi, M.; et al. Everolimus for the treatment of advanced, non-functional neuroendocrine tumours of the lung or gastrointestinal tract (RADIANT-4): A randomised, placebo-controlled, phase 3 study. *Lancet* **2016**, *387*, 968–977. [CrossRef]
162. Lacombe, C.; De Ryke, O.; Couvelard, A.; Turpin, A.; Cazes, A.; Hentic, O.; Gounant, V.; Zalcman, G.; Ruszniewski, P.; Cros, J.; et al. Biomarkers of Response to Etoposide-Platinum Chemotherapy in Patients with Grade 3 Neuroendocrine Neoplasms. *Cancers* **2021**, *13*, 643. [CrossRef]
163. National Cancer Institute-Children's Oncology Group Pediatric MATCH Trial. Children's Oncology Group. Available online: https://childrensoncologygroup.org/pediatricmatch (accessed on 14 July 2022).
164. Sundquist, F.; Georgantzi, K.; Jarvis, K.B.; Brok, J.; Koskenvuo, M.; Rascon, J.; van Noesel, M.; Grybäck, P.; Nilsson, J.; Braat, A.; et al. A Phase II Trial of a Personalized, Dose-Intense Administration Schedule of (177)Lutetium-DOTATATE in Children With Primary Refractory or Relapsed High-Risk Neuroblastoma-LuDO-N. *Front Pediatr.* **2022**, *10*, 836230. [CrossRef] [PubMed]
165. Advanced Accelerator Applications. A Multicenter Open-Label Study to Evaluate Safety and Dosimetry of Lutathera in Adolescent Patients With Somatostatin Receptor Positive Gastroenteropancreatic Neuroendocrine (GEP-NET) Tumors, Pheochromocytoma and Paragangliomas (PPGL). NCT04711135. Available online: https://clinicaltrials.gov/ct2/show/NCT04711135 (accessed on 20 September 2022).
166. Institut Claudius Regaud. PHASE I Clinical Study: Safety Evaluation of Peptide Receptor Radionuclide Therapy (PRRT) With 177Lu-DOTA0-Tyr3-Octreotate for Refractory or Recurrent Metastatic Neuroblastoma Expressing Somatostatin Receptors. NCT03966651. Available online: https://www.clinicaltrials.gov/ct2/show/NCT03966651 (accessed on 20 September 2022).
167. Menda, Y.; O'Dorisio, M.S.; Kao, S.; Khanna, G.; Michael, S.; Connolly, M.; Babich, J.; O'Dorisio, T.; Bushnell, D.; Madsen, M. Phase I trial of 90Y-DOTATOC therapy in children and young adults with refractory solid tumors that express somatostatin receptors. *J. Nucl. Med.* **2010**, *51*, 1524–1531. [CrossRef] [PubMed]

168. Gains, J.E.; Moroz, V.; Aldridge, M.D.; Wan, S.; Wheatley, K.; Laidler, J.; Peet, C.; Bomanji, J.B.; Gaze, M.N. A phase IIa trial of molecular radiotherapy with 177-lutetium DOTATATE in children with primary refractory or relapsed high-risk neuroblastoma. *Eur. J. Nucl. Med. Mol. Imaging* **2020**, *47*, 2348–2357. [CrossRef]
169. Erdmann, F.; Frederiksen, L.E.; Bonaventure, A.; Mader, L.; Hasle, H.; Robison, L.L.; Winther, J.F. Childhood cancer: Survival, treatment modalities, late effects and improvements over time. *Cancer Epidemiol.* **2021**, *71*, 101733. [CrossRef] [PubMed]
170. Panek, M.; Szymczak, M.; Stepaniuk, M.; Górecki, W.; Gawłowska-Marciniak, A.; Wolak, P.; Zbyrad, D.; Rybkiewicz, M.; Chrobak, K.; Noparlik, R.; et al. Radical surgical treatment of neuroendocrine tumors of the appendix in children—A Polish multicenter study. *Arch. Med. Sci.* **2021**, *17*, 1128–1131. [CrossRef]
171. Ramsey, M.J.; Nadol, J.B., Jr.; Pilch, B.Z.; McKenna, M.J. Carcinoid tumor of the middle ear: Clinical features, recurrences, and metastases. *Laryngoscope* **2005**, *115*, 1660–1666. [CrossRef] [PubMed]
172. Ferguson, M.K.; Landreneau, R.J.; Hazelrigg, S.R.; Altorki, N.K.; Naunheim, K.S.; Zwischenberger, J.B.; Kent, M.; Yim, A.P. Long-term outcome after resection for bronchial carcinoid tumors. *Eur. J. Cardiothorac. Surg.* **2000**, *18*, 156–161. [CrossRef]
173. American Thyroid Association Guidelines Task Force; Kloos, R.T.; Eng, C.; Evans, D.B.; Francis, G.L.; Gagel, R.F.; Gharib, H.; Moley, J.F.; Pacini, F.; Ringel, M.D.; et al. American Thyroid Association Guidelines—Medullary thyroid cancer: Management guidelines of the American Thyroid Association. *Thyroid* **2009**, *19*, 565–612. [CrossRef] [PubMed]
174. Hamidi, O.; Young, W.F., Jr.; Iniguez-Ariza, N.M.; Kittah, N.E.; Gruber, L.; Bancos, C.; Tamhane, S.; Bancos, I. Malignant Pheochromocytoma and Paraganglioma: 272 Patients Over 55 Years. *J. Clin. Endocrinol. Metab.* **2017**, *102*, 3296–3305. [CrossRef] [PubMed]
175. Robison, L.L.; Armstrong, G.T.; Boice, J.D.; Chow, E.J.; Davies, S.M.; Donaldson, S.S.; Green, D.M.; Hammond, S.; Meadows, A.T.; Mertens, A.C.; et al. The Childhood Cancer Survivor Study: A National Cancer Institute-supported resource for outcome and intervention research. *J. Clin. Oncol.* **2009**, *27*, 2308–2318. [CrossRef] [PubMed]
176. Hawkins, M.M.; Lancashire, E.R.; Winter, D.L.; Frobisher, C.; Reulen, R.C.; Taylor, A.J.; Stevens, M.C.G.; Jenney, M. The British Childhood Cancer Survivor Study: Objectives, methods, population structure, response rates and initial descriptive information. *Pediatr. Blood Cancer* **2008**, *50*, 1018–1025. [CrossRef] [PubMed]
177. Friedman, D.L.; Whitton, J.; Leisenring, W.; Mertens, A.C.; Hammond, S.; Stovall, M.; Donaldson, S.S.; Meadows, A.T.; Robison, L.L.; Neglia, J.P. Subsequent neoplasms in 5-year survivors of childhood cancer: The Childhood Cancer Survivor Study. *J. Natl. Cancer Inst.* **2010**, *102*, 1083–1095. [CrossRef] [PubMed]
178. Gunnes, M.W.; Lie, R.T.; Bjørge, T.; Syse, A.; Ruud, E.; Wesenberg, F.; Moster, D. Economic independence in survivors of cancer diagnosed at a young age: A Norwegian national cohort study. *Cancer* **2016**, *122*, 3873–3882. [CrossRef]
179. Brinkman, T.M.; Recklitis, C.J.; Michel, G.; Grootenhuis, M.A.; Klosky, J.L. Psychological Symptoms, Social Outcomes, Socioeconomic Attainment, and Health Behaviors Among Survivors of Childhood Cancer: Current State of the Literature. *J. Clin. Oncol.* **2018**, *36*, 2190–2197. [CrossRef]
180. Crochet, E.; Tyc, V.L.; Wang, M.; Srivastava, D.K.; Van Sickle, K.; Nathan, P.C.; Leisenring, W.; Gibson, T.M.; Armstrong, G.T.; Krull, K. Posttraumatic stress as a contributor to behavioral health outcomes and healthcare utilization in adult survivors of childhood cancer: A report from the Childhood Cancer Survivor Study. *J. Cancer Surviv.* **2019**, *13*, 981–992. [CrossRef] [PubMed]
181. Chauhan, A.; Kohn, E.; Del Rivero, J. Neuroendocrine Tumors-Less Well Known, Often Misunderstood, and Rapidly Growing in Incidence. *JAMA Oncol.* **2020**, *6*, 21–22. [CrossRef] [PubMed]

cancers

Review

Treatment of Gastrointestinal Stromal Tumors (GISTs): A Focus on Younger Patients

Monika Dudzisz-Śledź *, Anna Klimczak, Elżbieta Bylina and Piotr Rutkowski

Department of Soft Tissue/Bone Sarcoma and Melanoma, Maria Sklodowska-Curie National Research Institute of Oncology, Roentgena 5, 02-781 Warsaw, Poland; anna.klimczak@pib-nio.pl (A.K.); elzbieta.bylina@pib-nio.pl (E.B.); piotr.rutkowski@pib-nio.pl (P.R.)
* Correspondence: monika.dudzisz-sledz@pib-nio.pl

Simple Summary: Gastrointestinal stromal tumors (GISTs) are the most common mesenchymal neoplasms of the gastrointestinal tract. GISTs mainly develop in older adults, and the median age of diagnosis is 60–65 years. The incidence of GISTs in young adult patients, defined as adults before 40, is less than 10%. The frequency and type of molecular abnormalities in this group of patients are different from those in older patients. In this publication, we focus on the specificity of GISTs in young people and the principles of therapeutic management and management of the side effects of treatment.

Abstract: Gastrointestinal stromal tumors (GISTs) originate from Cajal's cells and are the most common mesenchymal neoplasms of the gastrointestinal tract. GISTs in young adults, i.e., patients before the age of 40, are rare and differ from those in older patients and GISTs in children in terms of the molecular and clinical features, including the location and type of mutations. They often harbor other molecular abnormalities than *KIT* and *PDGFRA* mutations (wild-type GISTs). The general principles of therapeutic management in young patients are the same as in the elderly. Considering some differences in molecular abnormalities, molecular testing should be the standard procedure to allow appropriate systemic therapy if needed. The optimal treatment strategy should be established by a multidisciplinary team experienced in sarcoma treatment. The impact of treatment on the quality of life and daily activities, including the impact on work, pregnancy, and fertility, in this patient population should be especially taken into consideration.

Keywords: gastrointestinal tumors; GIST; young adult; TKI; tyrosine kinase inhibitor; wild-type; KIT; PDGFRA; NF1; SDHB; SDH-competent; SDH-deficient

Citation: Dudzisz-Śledź, M.; Klimczak, A.; Bylina, E.; Rutkowski, P. Treatment of Gastrointestinal Stromal Tumors (GISTs): A Focus on Younger Patients. *Cancers* **2022**, *14*, 2831. https://doi.org/10.3390/cancers14122831

Academic Editors: Savio George Barreto and Irit Ben Aharon

Received: 21 May 2022
Accepted: 6 June 2022
Published: 8 June 2022

1. Introduction

Gastrointestinal stromal tumors (GISTs) usually develop in older people, and the median age of diagnosis is 60–65 years. GISTs rarely develop in younger patients. In children, GISTs often occur in girls, are located in the stomach, and generally do not have *KIT/PDGFR* mutations. The typical phenotypic and genotypic patterns in young adults aged 18 to 40 years are unknown. Less than 10% of GISTs are diagnosed in young people, i.e., before 40, and less than 1% of GISTs are diagnosed below 21 [1]. This disease in this population must be appropriately managed to optimize the efficacy and tolerability, primarily due to the long-expected survival and active participation in social and family life and the disease's impact on work, psychological aspects, and fertility. Young adulthood is a period of significant physical and psychosocial change, including continuing education, gaining financial independence, entering romantic relationships, starting a family, and raising children [2]. The disease and its treatment can interfere with daily activities and make it difficult to carry out daily activities. The treatment strategy should be defined by a multidisciplinary team experienced in soft tissue sarcomas, comprising an oncological surgeon, medical oncologist, pathologist, radiologist, gastroenterologist, and nuclear medicine specialists. Surgical treatment with R0 resection (negative margins), if possible, remains the mainstay of GIST management.

In some cases, preoperative therapy may be introduced. High- and intermediate-risk GISTs require adjuvant therapy [3]. For metastatic disease, targeted therapies are available, but surgery may also be used in some cases. Due to the molecular characteristics of young adult GISTs, specific considerations regarding appropriate therapy are needed to introduce the optimal treatment in young individuals with GISTs.

2. Epidemiology

GISTs are the most common mesenchymal neoplasms of the gastrointestinal tract. The incidence has been increasing during the last decades and, in most published studies, is reported at 10–15 new cases/per 100,000 per year [4–7].

3. Biology and Molecular Biology

The most common location of GISTs is the stomach (~60%). Less often, they are located in other parts of the gastrointestinal tract: about 30% of GISTs are detected in the jejunum and ileum, 5% in the duodenum, 5% in the rectum, and <1% in the esophagus [1]. Most GISTs are detected due to symptoms, but some are diagnosed as incidental findings during surgery or autopsy. The median size of GISTs ranges from 2.7 to 8.9 cm [5]. An example of advanced wild-type GIST originating in the stomach in young adult women treated for 18 years is shown in Figure 1.

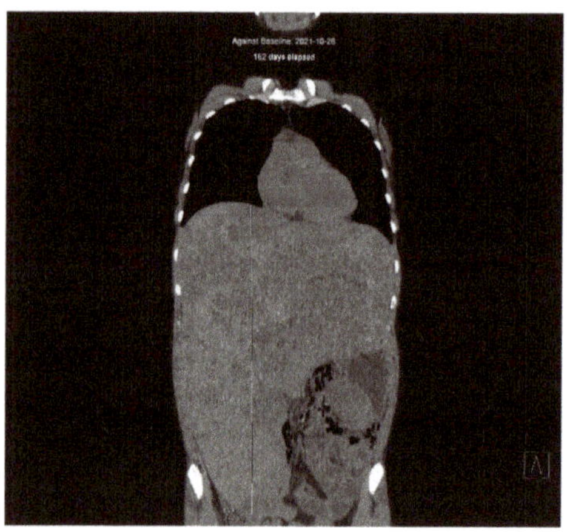

Figure 1. Advanced wild-type GIST originating in the stomach in young adult women treated for 18 years.

Two typical histological patterns of GIST are known. One is the spindle cell (60–70%) pattern and the second is epithelioid (30–40%). In some cases, a combination of both patterns in variable proportions occurs. GISTs stain positive for KIT (CD117, 95%) and DOG1 (almost exclusively characteristic for GIST). DOG1 expression is independent of the KIT status. About 5% of GISTs are CD117 negative, mainly in patients with the *PDGFRA* mutation [8].

Abraham et al. studied 150 esophagogastric resections for esophageal or esophagogastric junction carcinomas, and they found incidental GISTs in 15 of 150 (10%) patients [9]. All detected GISTs had a spindle cell morphology and were positive for CD117 and CD34. A study conducted in Germany showed a 25% incidence of small GISTs in the stomach in autopsies [10]. The majority of GISTs are sporadic and solitary, and about 80% of GISTs harbor activating mutations in the *KIT* or *PDGFRA* genes. These genes are responsible for the upregulation of crucial signaling pathways, including MAPK and PI3K-AKT. Up to 20% of CD117 (+) GISTs do not harbor *KIT* and *PDGFRA* mutations, i.e., are wild-type

GISTs [11]. *PDGFRA*-mutant tumors are primarily located in the stomach, mesentery, and omentum. GISTs with *KIT* exon 9 mutations primarily develop in the small intestine [12]. Most GISTs in children are wild type.

GISTs in young adults differ from pediatric GISTs and are not similar to typical GISTs in older patients. The data about the biology of GISTs in young adults is limited. IJzerman et al. published data from the Dutch GIST Registry for young adults aged 18 to 40 years and compared this data with data from older patients (>40). Of 1010 patients, 52 patients were ≤40 years. The authors found statistically significant differences between young and older GIST patients regarding the localization, mutational status, and presentation. The tumors were primarily located in the stomach (46%) and small intestine (46%). *KIT* mutations were diagnosed in 69% of patients and *PDGFRA* in 6%. In total, 25% of patients did not harbor *KIT* and *PDGFRA* mutations. Among them, 8% had *SDH*-deficient disease, 4% associated with NF1, 2% with *ETV6-NTRK3* gene fusion, and 10% were wild type. Young patients with GISTs were more often diagnosed in an emergency setting (18% vs. 9%). The overall 5-year survival rate was 85% [13].

Kang et al. analyzed 22 cases of GISTs in children and young adults up to 30 years of age. Of the 20 GISTs in young adults, 60% were located outside the stomach, and *KIT* or *PDGFRA* gene mutations were identified in 78% of the 18 cases. Ninety patients underwent R0 resection. One patient with a GIST located in the small intestine and with the *KIT* exon 11 deletion mutation had recurrent disease and was treated with imatinib with partial response. At diagnosis, one patient with multiple GISTs located in the stomach and with perigastric lymph node metastases developed multiple distant metastases and died after 7.3 years [14].

GISTs in young adults may have GIST features of child or adult GISTs. Prakars et al. analyzed the clinicopathologic and molecular features of 15 cases of GISTs in children and young adults (<30 years old). They included 15 patients with GIST, 5 children and 10 adults. Half of the 10 GISTs in young adults occurred in the small intestine and had a spindle cell morphology. In one case, lymph node metastasis was found. *KIT* mutations were identified in seven cases, four in exon 11 and three in exon 9. Recurrence was observed in seven patients [15].

Advances in molecular biology have allowed recognition that GISTs without *KIT/PDGFRA* mutations are usually deficient in succinyl dehydrogenase (SDH) due to silencing the epigenetic *SDHC* gene, and/or they have mutations in *NF1* and *BRAF* V600E, or *NTRK* gene rearrangement [16,17].

SDH mutations are more frequent in younger patients, especially in GISTs arising from the stomach. SDH-deficient GISTs make up 5% to 7.5% of all GISTs. The GIST with *SDH* mutations tends to metastasize, may metastasize to lymph nodes (which is unusual in typical GISTs), less frequently metastasize to the liver, usually grow slowly, and are often resistant to imatinib. In histological examination, these tumors are characterized by a multinodular growth pattern with epithelioid cells, which are multifocal. Additionally, it was found that SDH-deficient GISTs overexpress insulin-like growth factor receptors (IGF1R) [18]. Testing for germline mutations in *SDH* should be considered in young adults with GISTs wild type for *KIT/PDGFRA* mutations.

Some clinical syndromes are associated with GISTs, and these cases are usually diagnosed in young adults. One of them is the Carney's triad, which is usually a sporadic association of pulmonary chondroma, GISTs, and paraganglioma. GISTs in Carney's triad differ clinically, pathologically, and behaviorally from sporadic GISTs. Zhang et al. studied the clinical and pathologic features of the gastric neoplasm in 104 patients with Carney's triad. They found that GISTs in Carney's triad mainly affect young (mean age 22) women (88%), are often multifocal with higher epithelioid cell predominance, more often metastasize to lymph nodes, and often relapse. Their behavior is unpredictable [19]. The usual presentation was gastric bleeding. The second one, Carney-Stratakis syndrome, is a combination of familial paraganglioma and GISTs and is usually inherited and not associated with pulmonary chondroma. It is difficult to distinguish between Carney's triad and Carney-Stratakis syndrome due to the rarity of the components, and molecular testing may be helpful. GISTs in Carney-Stratakis syndrome and Carney's triad may be SDH-deficient GISTs. Carney's triad is usually caused by a specific pattern of methylation of the *SDHC* gene and may be due to germline mosaicism of the responsible genetic defect. Carney-Stratakis syndrome is instead caused by inactivating germline mutations in genes encoding for the SDH subunits [18].

In addition, type 1 neurofibromatosis (NF1), resulting from a loss-of-function mutation in the *NF1* gene, may be related to multifocal GIST, predominantly located in the jejunum or ileum, with the frequent presentation of gastrointestinal bleeding and anemia [20].

Based on the information above, primary resistance to imatinib is more common in GISTs in young populations due to the presence of mutations that prevent the molecule from binding to its KIT- and PDGFRA-binding sites. Wild-type GISTs are commonly insensitive to standard therapies, including imatinib [21–23]. About 10–15% of patients develop primary and early resistance during the first six months of therapy. The other most common genetic abnormality associated with primary resistance to imatinib is the D842V mutation, which relies on substitution of aspartic acid in codon 842 of PDGFRA into valine. Moreover, GISTs with mutations in exon 9 show a lower response rate and progression-free survival to imatinib than the most common exon 11 mutations when imatinib is used at a dose of 400 mg/day. In such cases, imatinib used at a dose of 800 mg/day is associated with better progression-free survival.

In addition to primary resistance, secondary resistance to imatinib may occur. The disease progresses in approximately 40–50% of patients during the first 2–3 years of imatinib therapy. This secondary resistance may be due to the accumulation of secondary point mutations in different regions of the *KIT* and *PDGFRA* genes [24]. It may also be due to the fibroblast growth factor (FGF) and the FGF receptor (FGFR) [25]. It has been shown that crosstalk between KIT and FGFR can promote imatinib resistance by reactivating the MAPK signaling pathway. Imatinib resistance may be promoted by crosstalk between KIT and FGFR due to the reactivation of the MAPK signaling pathway.

Hostein et al. analyzed a series of 321 GISTs for *BRAF* mutations and B-raf expression. They analyzed 251 GISTs with *KIT* or *PDGFRA* mutations and 70 wild-type GISTs. Among GISTs with *KIT* and *PDGFRA* mutations, no V600E mutation was detected. In wild-type GISTs, nine cases were positive for V600E mutation. GISTs with *BRAF* mutations were mainly localized in the small intestine and the stomach. No statistical difference in tumor location and other histologic and clinical features, including age, was found between WT GISTs with or without *BRAF* mutations. Three patients with *BRAF* mutations were high risk, three intermediate, two low, and in one case, the risk was not determined. They assessed BRAF expression in 37 GISTs (8 wild-type *BRAF* V600E-positive, 9 wild-type *BRAF*-negative, and 20 *KIT*- or *PDGRFA*-positive. BRAF expression was present in all cases. About 13% of *KIT* and *PDGFRA* wild-type GISTs are *BRAF* mutated [26].

Brenca et al. identified one fusion gene, *ETV6-NTRK3*, in one case of GIST among five *KIT/PDGFRA/BRAF* mutation-negative SDH-proficient tumors [27]. Shi et al. performed genetic comprehensive genomic profiling for coding regions in about 300 cancer-related genes of 186 GISTs to identify their somatic alterations. They found 24 GISTs without *KIT*, *PDGFRA*, and *RAS* mutations. Twelve did not harbor *SDH* alternations. The median age of patients with wild-type tumors was 44.4 years. The authors identified the most common mutated genes: *ARID1B*, *ATR*, *FGFR1*, *LTK*, *PARK2*, *SUFU*, and *ZNF217*. In two GISTs, *FGFR1* gene fusions were detected (*FGFR1–HOOK3*, *FGFR1–TACC1*), and one *ETV6–NTRK3* fusion that responded to TRK inhibition [28].

The summary of the main characteristics of GISTs in young people is presented in Table 1.

Table 1. Summary of the main characteristics of GISTs in young people.

Characteristics	
clinical	• are rare • more often develop in the stomach and small intestine • are more often diagnosed in emergency settings • the primary resistance to imatinib is more common • GISTs with *SDH* mutations tend to metastasize, including lymph nodes, less frequently metastasize to the liver, usually grow slowly, and are often resistant to imatinib • may be related to hereditary syndromes, such as Carney's triad or neurofibromatosis type 1
pathological	• may have GIST features of children or adults GIST
molecular	• more frequent wild type, more frequent mutations related to resistance to imatinib, including SDH mutations

4. Treatment

4.1. Surgery and Perioperative Therapy

The young adult population is a particular group of patients. Their lifestyle, lack of routinely used screening tests, delays in presentation, lower incidence of neoplasms, and, therefore, less frequent association of symptoms with suspected neoplasms may delay diagnosis and treatment.

The data on the treatment of GISTs in young patients is limited. As with all patients with GISTs, the main goal of GIST treatment is the surgical removal of tumors with histologically negative margins (R0). Molecular testing is recommended.

It was shown, based on the National Cancer Institute's Surveillance, Epidemiology, and End Results (SEER) database analysis of patients with GISTs, that surgical treatment in people below 40 is performed more often (84.7% vs. 78.4%, $p = 0.003$), and is related to better outcomes in terms of GIST-specific survival (GSS) and overall survival (OS) in comparison to patients ≥40, including patients with metastases. This retrospective cohort study included 392 young patients (13–39) and 5373 older adult patients diagnosed from January 2001 until December 2013, with follow-up until December 2015. Young patients were more often diagnosed with small intestine GISTs than older patients (35.5% vs. 27.3%, $p = 0.008$). In the young patient subgroup with GISTs located in the stomach and small intestine, the small intestine location was associated with improved OS (91.1% vs. 77.2%, $p = 0.01$%), and GSS (91.8% vs. 78.0%, $p = 0.008$) and were more often treated surgically (84.7% vs. 78.4%, $p = 0.003$). In general, surgical treatment improved the prognosis in young patients with GISTs. Management without surgery was associated with a more than two-fold increased risk of death from GISTs [29].

Extended anatomic resections and complex multivisceral resections should be avoided whenever possible. In low-risk GISTs located in unfavorable locations, the R1 resection may be acceptable, and the decision should be made with the patient [30]. Usually, for R1 resection, routine re-excision is not recommended on a routine basis, and the microscopic margin status should not be taken when deciding on adjuvant therapy. If extended surgery is required, imatinib neoadjuvant therapy should be considered in GISTs with imatinib-sensitive mutations.

For GISTs insensitive to imatinib, i.e., GISTs with *PDGFRA* exon 18 mutations (including the D842V mutation), neoadjuvant avapritinib may be considered. Before neoadjuvant therapy, a biopsy should be performed to confirm the diagnosis and molecular testing. Imatinib can be continued in a preoperative setting until the maximum response, with close response assessment to avoid missing the resistance and progression. Imatinib should be continued in an adjuvant setting after surgery. Patients after preoperative avapritinib should undergo observation after surgery [30]. Patients with *PDGFRA* mutations and wild-type GIST after curative surgery have a lower risk of recurrence than patients with KIT mutations [3,30].

In patients who were treated systemically perioperatively, imatinib can be stopped right before surgery and restarted as soon as the patient can tolerate oral medications. Avapritinib should be stopped at least one week prior to surgery.

During the surgery, every effort should be made not to disturb the tumor pseudocapsule. One of the adverse prognostic factors is tumor rupture.

In some circumstances, a laparoscopic (favorable anatomic locations, small tumors) or endoscopic (small tumors in the upper or lower GI tract) approach may be considered. Based on a meta-analysis of 19 studies, including 1060 patients with GISTs, it was shown that there is no difference between laparoscopy and laparotomy regarding long-term outcomes. Laparoscopy was associated with less blood loss, lower complication rates, and shorter hospitalization [31].

The incidence of nodal metastases is low, and lymphadenectomy is usually not required. Lymphadenectomy must be considered in patients with known SDH-deficient GISTs or translocation-associated GISTs and pathologically enlarged nodes. For SDH-deficient GISTs with multifocal disease, extensive surgery associated with significant morbidities, such as total gastrectomy, is not recommended to reduce the risk of recurrence in the stomach [3]. In GIST patients with SDH deficiency or known *SDH* mutations, the risk of paraganglioma is increased, and diagnostic tests should be considered prior to surgery. Patients with *SDH* mutations are at risk of paraganglioma, and they should be tested using 24-h urine collection prior to the surgery [3].

The available data indicate that perioperative imatinib should be used for three years (including preoperative therapy). As per NCCN guidelines, adjuvant imatinib is preferred for patients with a significant risk of recurrence, i.e., intermediate or high risk. The ESMO guidelines recommend adjuvant therapy for high-risk patients, and for patients with intermediate risk, the decision should be individualized, and the decision-making process may include genotyping for KIT mutations [3,30].

Three years of therapy improved the relapse-free survival (RFS 65.6% vs. 47.9%) and overall survival (OS, 5-year OS: 92% vs. 81.7%, 10-year OS: 79% vs. 65.3%) in comparison to the one-year therapy [32,33]. The phase II PERSIST-5 study has shown that five years of adjuvant imatinib therapy was associated with little risk of recurrence in patients after resection of intermediate- or high-risk GISTs [34]. The 5-year recurrence-free survival was 90%, and the 5-year overall survival was 95%; 49% of patients did not complete therapy, with the time from treatment discontinuation to relapse ranging from 7 to 24 months [35]. The ESMO guidelines recommend perioperative therapy with imatinib for three years [30].

Patients with *PDGFRA* D842V-mutated GISTs should not be treated with adjuvant therapy due to the lack of imatinib efficacy and other systemic options in adjuvant settings [30]. Adjuvant treatment, including imatinib, is not recommended in NF1-related and SDH expression-negative, *BRAF*-mutated, or *NTRK*-rearranged GISTs [30].

GIST staging is usually based on the American Joint Committee on Cancer (AJCC) TNM classification system 8th edition from 2017 [3].

The risk assessment uses tumor features such as the primary mitotic count, tumor size, and tumor site, and standard risk classification does not include mutational status [36,37].

4.2. Treatment of Unresectable/Metastatic GISTs

In unresectable and metastatic settings, systemic therapy with targeted therapies is the mainstay of disease management. GISTs are generally resistant to chemotherapy and radiation therapy.

Medical therapy in young adults should be simplified to adjust to daily activities, whenever possible. The visits and consultations should be flexible to enable young adults to continue therapy without impacting school or work and daily activities. The patients should be well educated about the disease, prognosis, treatment options, and additional support, including mental health specialist and psychological counseling. It should also include counseling about fertility impairment and preservation before treatment initiation and its impact on sexual health. Smokers should be referred to a smoking cessation program. Patients should also be educated on dietary recommendations and possible changes related to cancer treatment [38].

Imatinib is the standard of care in the first line of unresectable and metastatic GISTs. It was the first drug introduced into clinical practice in GISTs. It is a KIT and PDGFRA tyrosine kinase inhibitor. Before discovering imatinib, the median OS of patients with unresectable or metastatic disease was 12–15 months. Based on prospective clinical trials, imatinib improved OS to approximately five years. The median PFS was 2–3 years. Complete responses were rarely observed in 5–7% of patients, but partial responses were observed in 40% of patients, and disease stabilization in 36% [39–41]. These data were confirmed in clinical practice [39,42]. Imatinib is used at a daily dose of 400 mg, and the daily dose may be increased in case of disease progression to 800 mg [43]. The best responses to imatinib occur in GISTs with mutations in *KIT* in exon 11. GISTs with exon 9 *KIT* mutations and GISTs without *KIT* mutations, and GISTs with specific mutations in the PDGFRA gene, especially D842V, are less or insensitive to imatinib.

For adult patients with unresectable or metastatic GIST harboring the *PDGFRA* D842V mutation, which is resistant to imatinib, avapritinib has been approved for first-line therapy based on phase I NAVIGATOR study results. Avapritinib is a type 1 kinase inhibitor that has demonstrated activity on the *PDGFRA* D842V and *KIT* D816V mutants associated with resistance to imatinib, sunitinib, and regorafenib. The USA approved the drug to treat unresectable or metastatic GIST patients harboring *PDGFRA* exon 18 mutations, including the *PDGFRA* D842V mutation. In Europe, this drug has been approved for *PDGFRA* D842V GISTs. In patients with GISTs with the *PDGFRA* D842V mutation treated with different doses, the objective response rate (ORR) was 88%, with CR in 9% of patients, PR in 79%, and SD in 13%. The ORR among 38 patients with the *PDGFRA* D842V mutation treated with avapritinib 300 or 400 mg was 95%, with CR in 13% of patients and PR in 82%. The median PFS was 24 months. Median OS was not reached, and the duration of response was 22 months [44,45].

In the phase III VOYAGER study, avapritinib was compared to regorafenib in patients with unresectable or metastatic GIST previously treated with imatinib and one or two other tyrosine kinase inhibitors (TKIs). The primary endpoint of an improvement in PFS was not met. PFS was 4.2 months in the avapritinib arm and 5.6 months in the regorafenib arm [46].

Patients who progress or are intolerant to imatinib may be treated with sunitinib in the second-line therapy. Sunitinib is a multitargeted TKI that targets KIT receptor tyrosine kinase, PDGFR, VEGFR, and FLT3. This is the only TKI approved for GIST therapy in the second line. Up to 40% of patients on sunitinib, especially with the exon 9 *KIT* mutation, may achieve long-term responses, with the time to progression 6–8 months (median). In a phase III, randomized, placebo-controlled, double-blind study, the progression-free survival was 22.9 vs. 6.0 weeks in the placebo arm. PFS and OS were longer in patients with a primary *KIT* exon 9 mutation or wild-type *KIT/PDGRFA*. The recommended dose of sunitinib is 50 mg taken orally once daily for 4 consecutive weeks, followed by a 2-week rest period (schedule 4/2). Dose modifications in 12.5 mg steps may be applied for toxicity management based on individual safety and tolerability. An option is a continuous dosing regimen with a dose of 37.5 mg daily without interruption [47–50].

Regorafenib is another oral inhibitor that potently blocks multiple protein kinases, including kinases involved in tumor angiogenesis (VEGFR1, -2, -3, TIE2), oncogenesis (KIT, RET, RAF-1, BRAF, BRAFV600E), metastasis (VEGFR3, PDGFR, FGFR), and tumor immunity (CSF1R) [51]. The drug is approved for treatment of adult patients with GISTs who have been previously treated with other anticancer medicines (imatinib and sunitinib). The recommended dose is 160 mg taken orally once daily for the first 21 days of each 28-day cycle. Regorafenib was assessed in a multicenter phase II study [52] and phase III GRID trial [53]. In the previously pretreated population, the mean PFS in the regorafenib group was more than five times longer than in the placebo group.

Ripretinib is a TKI that inhibits KIT proto-oncogene receptor tyrosine kinase and PDGFRA kinase, including wild-type, primary, and secondary mutations, and inhibits other kinases in vitro such as PDGFRB, TIE2, VEGFR2, and BRAF. Ripretinib was assessed, in comparison to a placebo, in the phase II INVICTUS study in patients with disease progression on at least imatinib, sunitinib, and regorafenib or who had documented intolerance to any of these medications despite dose modifications. Ripretinib, as a fourth or further treatment line, significantly improved median PFS compared with the placebo (6.3 vs. 1.0 months; HR 0.15; 95% CI 0.09–0.25; $p < 0.0001$). Median OS was 15.1 and 6.6 months in the ripretinib and placebo arms, respectively (HR 0.36; 95% CI 0.21–0.62) [54].

Ripretinib has also been assessed compared to sunitinib in the phase III INTRIGUE study in patients with advanced GIST after treatment with imatinib. PFS was not statistically different between the ripretinib and sunitinib arms, and median OS was not reached in either arm [55].

A summary of the efficacy of drugs approved for treatment of unresectable and metastatic GISTs is provided in Table 2.

Table 2. Summary of the efficacy of drugs approved for treatment of unresectable and metastatic GISTs.

Authors and Type of Study	Drug	mPFS	mOS	ORR (%)
Blanke et al. phase III randomized trial 2008 (NCT00009906) [56]	imatinib 400 mg vs. imatinib 800 mg	18 vs. 20 months	55 vs. 51 months	43 vs. 41
Demetri et al. phase III randomized trial 2006 (NCT00075218)	sunitinib vs. placebo	22.9 vs. 6.0 weeks	72.7 vs. 64.9 weeks	6.6 vs. 0
Demetri et al. [53] phase III randomized trial 2013 (NCT01271712)	regorafenib vs. placebo	4.8 vs. 0.9 months	HR 0.77; $p = 0.199$	75.9 vs. 34.8
Jean-Yves Blay et al. phase III randomized trial 2020 (NCT03353753) [54]	ripretinib	6.3 vs. 1.0 months	15.1 vs. 6.6 months	9
Jones et al. [45] phase I (NCT025085320)	avapritinib (data for patients with PDGFRA D842V mutation)	NR; PFS at 3 months 100%; 6 months 94%, 12 months 81%	NR; estimated OS at 6 months 100%, 12 months 91%, 24 months 81%	88

mPFS—median progression-free survival; mOS—median overall survival; ORR—objective response rate; NR—not reached; HR—hazard ratio; PDGFRA platelet-derived growth factor receptor A.

Pazopanib, an orally administered, potent multi-target tyrosine kinase inhibitor (TKI) of VEGFR1, VEGFR2, VEGFR3, PDGFRA, PDGFRB, and c-KIT, was assessed in the phase II PAZOGIST study in patients with GIST ($n = 81$) with failure on imatinib and sunitinib. The 4-month PFS rate in central assessment was significantly higher in the pazopanib group (44.3%) compared to the control group (17.6%). Based on the investigator's assessment, the median PFS was 3.4 months in the pazopanib arm and 2.3 months in the placebo group [57].

Sunitinib, regorafenib, and pazopanib may be more effective in GISTs with *SDH* mutations and SDH-deficient GISTs. The multicenter series of pediatric/young adult patients with advanced *KIT/PDGFRA* WT GISTs treated with sunitinib (potent antiangiogenic inhibitor) confirmed some clinical benefits of sunitinib in this population [58]. These data were similar to the series of Janeway et al. in pediatric GIST patients. A longer time to progression on sunitinib compared to prior imatinib therapy was observed [59].

Sorafenib, a multikinase inhibitor with activity against KIT and PDGFRA and several other kinases, was assessed in two single-arm phase II clinical trials in patients with GISTs after progression during therapy with imatinib and sunitinib. Median PFS in both trials was about 5 months, and OS was 9.7 and 11.6 months, respectively [60,61].

Patients with GIST with *NTRK* rearrangement may be sensitive to the neurotrophic tyrosine receptor kinase (NTRK) inhibitors larotrectinib and entrectinib. The efficacy of NTRK inhibitors was shown in clinical trials in patients with solid tumors [30,62,63].

Patients with GISTs with *BRAF* mutations may benefit from BRAF inhibitors (including the anti-BRAF plus anti-MEK combination) [30]. Falchook et al. showed the activity of the BRAF inhibitor dabrafenib in non-melanoma solid tumors in a phase II study [64]. They also showed dabrafenib's antitumor activity in GIST patients with the V600E *BRAF* mutation [65].

Other molecules that may be effective in some GISTs include nilotinib, ponatinib, dasatinib, cabozantinib, and crenolanib.

Nilotinib, a selective and potent TKI targeting BCR-ABL, c-KIT, PDGFR, and other kinases, is effective in patients who failed both imatinib and sunitinib due to disease progression or intolerance. In the study conducted by Montemurro et al. (n = 52), 10% of patients responded to nilotinib, and 37% achieved disease stabilization. Median PFS was 12 weeks, and median OS was 34 weeks [66]. In the post hoc subset analyses in the phase III study, nilotinib provided significantly longer median OS in patients pretreated with imatinib and sunitinib [67].

Ponatinib is a potent pan BCR-ABL inhibitor, which demonstrated activity against RET, FLT3, and KIT and members of the FGFR, PDGFR, and VEGFR families of kinases [68]. This drug has shown activity against the KIT exon 17 D816-mutant kinases [69]. In a phase II single-arm study (n = 42), the clinical benefit rate in patients with KIT exon 11 mutations at 16 weeks was 37% [70]. In the POETIG study, another phase II trial that used ponatinib at a reduced dose, the clinical benefit rate was 35%, and the median PFS was 86 days [71].

Dasatinib, a potent inhibitor of BCR-ABL, KIT, and SRC family kinases, was assessed in a phase II study in patients with GIST resistant to imatinib. The median PFS and OS were 2 and 19 months, respectively, and the median PFS in wild-type GISTs was 8.4 months [72]. As per current NCCN guidelines, dasatinib may be used for patients with *PDGFRA* exon 18 mutations insensitive to imatinib, including *PDGFRA* D842V mutations [3].

Cabozantinib, a multitargeted TKI targeting KIT, VEGFR-2, MET, and AXL, was assessed in patients with metastatic GIST after treatment with imatinib and sunitinib in the phase II CaboGIST study. The median PFS at 12 weeks was 60%. DCR was 82%, with PR in 14% and SD in 68% of patients. The median PFS was 5.5 months, and the median OS was 18.2 months [73].

Crenolanib is a TKI with activity against PDGFR and FLT3 with nanomolar activity against *PDGFRA* D842V mutant GIST. Crenolanib was used in a phase II study and is currently being assessed in a randomized, double-blinded, placebo-controlled phase III trial in advanced and metastatic PDGFRA D842V mutant GIST (CrenoGIST) [74–76].

Some molecular data indicate that O6-methylguanine-DNA methyltransferase (MGMT) promoter methylation is prevalent in SDH-deficient GISTs, suggesting sensitivity to alkylating agents. Yebra et al. presented data during the 2020 Annual Meeting of Connective Tissue Oncology Society, which demonstrated the therapeutic vulnerability of SDH-deficient GISTs to the DNA alkylating agent, temozolomide, and a 40% rate of objective responses among five patients treated with this drug. The disease control rate was 100% [77,78]. A phase II study (NCT03556384) is ongoing. Further preclinical and clinical research on SDH-deficient GISTs is needed.

For patients with metastatic GISTs with mutations resistant to imatinib, sunitinib, regorafenib, ripretinib, and avapritinib, referral to a clinical trial is recommended [3].

In patients with limited progression, when standard and investigational therapies fail, re-challenge with a TKI that was previously tolerated and effective for symptom palliation may be considered. Patients who progressed on TKI may experience tumor growth acceleration after TKI discontinuation. In a phase III RIGHT study (n = 81), after a median follow-up of 5.2 months, the median PFS was 1.8 months in the imatinib group and 0.9 months in the placebo arm (HR 0.46, 95% CI 0.27–0.78; p = 0.005) [79].

Surgery may be performed in unresectable and metastatic disease, such as limited disease progression in GIST refractory to imatinib, symptomatic bleeding or obstruction, locally advanced or previously unresectable disease, or low-volume metastatic disease after response to imatinib. Imatinib can be stopped just before surgery and restarted as soon as the patient tolerates oral medications. If the patient is treated with other TKIs than imatinib, such as regorafenib, sunitinib, avapritinib, or ripretinib, the medication should be stopped at least one week prior to surgery and can be restarted based on the clinical assessment [3]. Based on the National Cancer Institute's Surveillance, Epidemiology, and End Results (SEER) database analysis of patients with GISTs, operative management in young adults with metastases was associated with improved OS (69.5% vs. 53.7%, $p = 0.04$) and GSS (71.5% vs. 56.7%, $p = 0.03$) [29]. The Figure 2 shows CT scans young woman with GIST with KIT exon 9 deletion mutation arising from the small intestine, after treatment with TKIs, who undergo surgical resection of residual lesion in abdominal cavity.

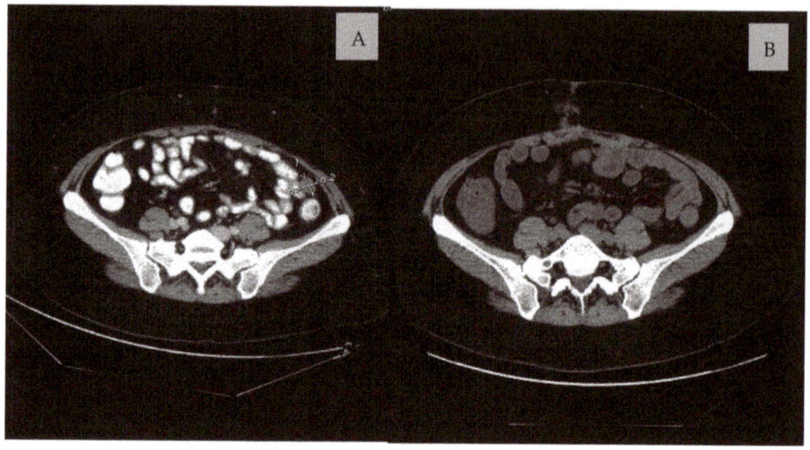

Figure 2. CT scan before (**A**) and after (**B**) resection of residual metastatic lesion in the abdominal cavity in young women with GIST with *KIT* exon 9 deletion mutation arising from the small intestine, after treatment with TKIs.

For patients with SDH mutations and SDH deficiency and patients with NF1 mutations, genetic counseling should be advised [3].

5. Treatment Tolerability

There are no specific data on the tolerability of drugs used in the treatment of GIST in young adults. Young adults often have overall better organ function than older adults and have fewer comorbidities, take fewer medications, and, therefore, are at lower risk of drug–drug interactions. They may tolerate treatment better than older people. For very young adults, puberty is also a time of significant physical activity and increased physiological changes in body composition, protein binding, and organ function, all of which can affect drug metabolism [38]. Due to the lack of detailed information on the treatment tolerability of individual drugs used in patients with GISTs, the overall data on the toxicity of therapy are summarized below. The most common toxicities of the most frequently used medications are summarized in Table 3.

Table 3. Adverse events reported in clinical trials with drugs approved for the treatment of unresectable and metastatic GISTs.

Authors and Type of Study	Drug	Frequency of Drug Related AEs		Most Frequent Drug-Related AEs		SAEs		AEs Leading to Treatment Discontinuation	Frequency of Dose Modifications
		All Grades	Grade at Least 3	All Grades	Grade at Least 3	Any Grade (Frequency)	Most Frequent		
Blanke et al. [56] phase III randomized trial 2008 (NCT00009906)	imatinib	NA	400 mg 43% 800 mg 63%	NA	400 mg: anemia 9%, GI toxicities 9%, neutropenia 7%, cardiac toxicities 6%, hemorrhage 5% 800 mg: GI toxicities 16%, anemia 14%, cardiac toxicities 14%, neutropenia 10%, hemorrhage 11%	NA	NA	Most common 400 mg: rash, edema, GI bleeding 800 mg: edema, nausea, fatigue	400 mg: at least one dose delay 38%; at least one dose reduction 16% 800 mg: at least one dose delay 59%; at least dose reduction 58%
Demetri et al. [47] phase III randomized trial 2006 (NCT00075218)	sunitinib	83%	NA	anemia 62%, neutropenia 53%, thrombocytopenia 41%, fatigue 34%, diarrhea 29%, skin discoloration 25%, nausea 24%	neutropenia 10%, thrombocytopenia 5%, fatigue 5%, anemia 4%, HFS 4%, diarrhea 3%, asthenia 3%, hypertension 3%,	20%	HFS, diarrhea, hypertension	7.2%	NA
Demetri et al. [53] phase III randomized trial 2013 (NCT01271712)	regorafenib	98%	61%	HFS 56%, hypertension 49%, diarrhea 40%	hypertension 23%, HFS 20%, diarrhea 5%	29%	abdominal pain 4%, fever 2%, dehydration 2%	6%	72%
Jean-Yves Blay et al. [54] phase III randomized trial 2020 (NCT03353753)	ripretinib	NA	NA	alopecia 49%, myalgia 28%, nausea 26%, fatigue 26%, HFS 21%, diarrhea 21%	lipase increased 5%, hypertension 4%, fatigue 2%, hypophosphatemia 2%	9%	All SAEs: anemia, cardiac failure, death of unknown cause, dyspnea, fecaloma, GERD, hyperkalemia, hypophosphatemia, nausea, upper GI hemorrhage	5%	NA
Jones et al. [45] phase I (NCT025085320)	avapritinib 300 mg/d	99%	65%	nausea 69%, anemia 56%, diarrhea 47%, fatigue 41%, decreased appetite 38%, periorbital edema 37%	anemia 22%, neutropenia 9%, decreased neutrophile count (9% diarrhea 6%)	26% in safety population	Anemia, pleural effusion, diarrhea, vertigo	12% in safety population, 14% in D842V population	NA

AE—adverse event, GERD—gastroesophageal reflux disease, GI—gastrointestinal, HFS—hand foot skin reaction, NA—not available, SAE—serious AE.

5.1. Imatinib

Drug–drug interactions, compliance, and the genetic variability of metabolizing or drug-resistant enzymes impact the drug concentrations and, as a result, the efficacy and tolerability of imatinib. In adults treated with imatinib, no significant age-related pharmacokinetic differences were observed [80]. The imatinib-related adverse events (TRAE, treatment-related adverse event), hematological and non-hematological, are well known. The most common adverse events (AEs) include fluid retention and edemas (particularly periorbital), abdominal pain, diarrhea, nausea, vomiting, fatigue, rash, and muscle cramps in the fingers and feet. SAEs, such as liver function test abnormalities, lung toxicity, gastrointestinal bleeding, and hematological AEs, have been reported rarely [3].

Side effects may improve with prolonged therapy and can usually be managed with appropriate supportive treatment, but some patients need to discontinue or reduce the dose of the drug. In some studies, imatinib discontinuation has been associated with rapid disease progression. The NCCN GIST guidelines recommend continuing imatinib treatment at a reduced dose if an adverse effect recurs after discontinuation [3]. In the PERSIST-5 study with 5 years of adjuvant imatinib, only 51% completed 5 years of imatinib therapy, and 16% discontinued therapy due to adverse events [35]. If grade 3 neutropenia or thrombocytopenia occurs, the drug should be discontinued until improvement to at least grade 1 (neutrophils > 1.5×10^9/L; platelets > 75×10^9/L) is achieved. Administration of the drug may be resumed at the dose used prior to the adverse event. If the event recurs, the drug should be discontinued and resumed at a lower dose [43]. Cases of acute liver damage (acute hepatitis) have been reported. Liver function should be monitored regularly in patients treated with imatinib. In such cases, prednisolone appears to be useful [80]. The minimum recommended dose of 400 mg of imatinib per day should be used in patients with hepatic impairment. Patients should be aware of this potential complication and know the factors that may increase the risk, such as drug–drug interactions and the effect of certain foods, including alcohol. An increase in the bilirubin concentration to >3 × ULN or an increase in liver transaminases to >5 × ULN requires discontinuation of the drug until bilirubin levels return to <1.5 × ULN and transaminase levels <2.5 × ULN. The patient may continue therapy in reduced doses (from 400 to 300 mg daily, from 600 to 400 mg daily, or from 800 to 600 mg daily) [81]. During treatment with imatinib, other drugs, including protease inhibitors, azole antifungals, selected macrolides, CYP3A4 substrates with a narrow therapeutic window, warfarin, and other coumarin derivatives, should be taken with caution. The patients should avoid the consumption of grapefruit and grapefruit juice and avoid the potent inhibitors of CYP3A. Caffeine may increase the activity of imatinib, and, therefore, caffeine-containing products should not be consumed during treatment with imatinib [81]. Agents that induce CYP3A4, such as carbamazepine, dexamethasone, phenobarbital, phenytoin, rifampicin, and Hypericum perforatum, which may reduce exposure to imatinib and make it less effective, should be avoided [43].

5.2. Sunitinib

In 2011, Hutson et al. [82] published the pooled data of 1059 patients who received sunitinib 50 mg/day on an approved regimen of 4 weeks every 2 weeks (n = 689) or a continuous dose of 37.5 mg once daily (n = 370). A total of 857 (81%) patients were aged <70 years, and 202 (19%) were aged ≥70 years. The median age for the <70 years group was 57 (range 24–69). Treatment tolerability was similar in both groups. Most treatment-related adverse reactions occurred with a similar frequency in both age groups. Adverse effects that occurred significantly less frequently in younger patients were decreased appetite/weight loss, cough, fatigue, edema, and anemia. An AE that occurred more frequently in younger patients was hand-foot syndrome [82]. Hematologic AEs reported during treatment with sunitinib include anemia, thrombocytopenia, and neutropenia. The most commonly reported gastrointestinal AEs were diarrhea, nausea/vomiting, abdominal pain, dyspepsia, and oral ulceration. Treatment of gastrointestinal AEs depends on the severity and includes antiemetic and antidiarrheal drugs [83]. In case of grade 3 diarrhea, sunitinib should be interrupted until improvement to grade 1, and treatment should be resumed at a reduced dose [83,84]. AEs associated with sunitinib treatment often lead to a dose reduction or temporary interruption [84]. The specific AE that may occur during therapy with sunitinib is hypothyroidism, which often does not require treatment discontinuation, and thyroid hormone replacement therapy is sufficient [83,84]. In patients scheduled to undergo major surgery, temporary discontinuation of sunitinib treatment is recommended due to the impaired wound healing observed during sunitinib treatment [83]. Another AE reported by patients is oral mucositis. In the case of grade 3 and 4 mucositis, it is recommended to temporarily stop therapy and reinstitute it with a dose reduction after improvement [84,85]. The next AE often reported by patients treated with sunitinib is hypertension, which should be treated appropriately with antihypertensives. If severe hypertension cannot be controlled with available medications, it may be necessary to discontinue sunitinib treatment temporarily. Treatment may be resumed once hypertension is adequately controlled [83,84]. Cardiac events such as heart failure, myocarditis, decreased left ventricular ejection fraction, cardiomyopathy, and myocardial infarction have also been reported with sunitinib treatment. Treatment with sunitinib should be discontinued if clinical signs or symptoms of heart failure appear [83]. HFS is another AE reported by patients treated with sunitinib, and is most frequently reported as grade 1–2 in 13% and grade 3–4 in 4% of sunitinib-treated patients [86,87]. In order to prevent the occurrence of HFS, patients should be advised to use moisturizing creams from the beginning of sunitinib treatment [84]. Grade 3–4 skin rash is relatively rare but may require temporary interruption or a dose reduction. Treatment, including the use of topical steroids, is recommended. Grade 1–2 skin rash occurs in 14% of sunitinib-treated patients. Skin discoloration has been observed in 30% of sunitinib-treated patients, and alopecia is also an AE reported by sunitinib-treated patients [86,87].

5.3. *Regorafenib*

The safety and efficacy of regorafenib in the treatment of GIST were evaluated in the Phase III GRID trial, where 98% of patients reported AEs of any grade and 72% of patients required dose modification [88]. Treatment termination due to AEs affected 40% of patients treated with regorafenib [53]. These data are similar to a phase II trial of regorafenib in GIST, where 82% of patients required dose modification [52], and to the results of a retrospective analysis of data from 50 GIST patients treated with regorafenib presented by Chamberlain et al. [88]. The most frequent grade 3–4 AEs were HFS and fatigue. The general tolerability profile of regorafenib is quite similar to sorafenib. HFSR (hand-foot skin reactions), hypertension, diarrhea, and fatigue are the most commonly reported AEs observed in clinical trials with both drugs [89]. Treatment of HFSR may include oral or topical analgesics and cool compresses on the skin in grade 1, and topical therapy and oral steroids and/or anesthetics are recommended in grade 2 and 3. Fatigue should be managed by deficiency corrections (vitamin D3, anemia), lifestyle changes (physical exercise, sleep hygiene), nutritional support, or regorafenib dose modification [89].

5.4. *Avapritinib*

In a phase III clinical study with avapritinib in GISTs with exon 18 PDGFRA mutation, the most common AEs reported in at least 10% of patients were edema, nausea, vomiting, decreased appetite, fatigue/asthenia, and cognitive impairment. SAEs reported in this study included anemia (9%), abdominal pain (3%), pleural frostbite (3%), sepsis (3%), gastrointestinal hemorrhage (2%), acute kidney injury (2%), vomiting (2%), pneumonia (1%), and tumor hemorrhage (1%). AEs with fatal outcomes were reported in 3.4% of patients. In total, 49% of patients required dose modifications (reduction or treatment discontinuation). In total, 48% of participants treated with avapritinib experienced grade 1–3 cognitive impairment [90].

5.5. Ripretinib

The tolerability of ripretinib was assessed in the phase III INVICTUS study. The most common TEAEs reported in this study in the ripretinib group were alopecia, nausea, diarrhea, fatigue, myalgia, and HFS. The most common treatment-related grade 3 and 4 events were hypertension, fatigue, hypophosphatemia, and increased lipase. Treatment-related SAEs were observed in 9% of patients who received ripretinib and included anemia, heart failure, dyspnea, gastroesophageal reflux disease, and death of unknown cause. TRAEs leading to dose reductions and treatment discontinuation occurred in 6% and 5% of participants. The safety profile of ripretinib was acceptable [54].

5.6. Sorafenib

The most common adverse reactions reported in the phase II study conducted by the Korean GIST Study Group were: HFS, skin rash, abdominal pain, and diarrhea; these were grade 1 and 2 adverse reactions, and most of them were reversible. Ten patients required a dose reduction or treatment discontinuation due to intolerance. The most common adverse reactions resulting in dose reductions were: HFS, rash, hypertension, and diarrhea. No toxicity-related deaths were reported. No toxicity-related deaths were observed [60]. Similar grade 3–4 adverse events were also observed in the phase II study published by Kindler et al. [61], including HFS, hypertension, diarrhea, hypophosphatemia, gastrointestinal bleeding, gastrointestinal perforation, thrombosis, and intracranial hemorrhage, and 61% of patients required dose reductions [61].

5.7. Pazopanib

In the PAZOGIST study, AEs with grades of at least 3 were observed in 72% of participants in the pazopanib group vs. 17% in the control arm. The most frequently reported AE was hypertension, reported by 38% of the participants. In total, 26% of the patients reported treatment-related SAEs [57].

5.8. Dasatinib

The most common AEs of dasatinib observed in a prospective phase II study by Zhou et al. were anemia, proteinuria, fatigue, neutropenia, and diarrhea. In total, 6.9% of the participants discontinued dasatinib due to AEs before the first efficacy assessment, and 17.2% of the patients reported grade 1 gastrointestinal bleeding during treatment [91]. Fluid retention events such as pleural effusion in all grades and grade 3 and 4 occurred in 13% and 6%, respectively [92].

5.9. Cabozantinib

The most common TRAEs reported in the CaboGIST study were diarrhea, fatigue, hypertension, stomatitis, weight loss, and HFS. Overall, tolerability was similar to that of other TKIs and was controlled by dose modification and supportive care [73].

5.10. Ponatinib

The most common grade 3 and 4 AEs observed in the POETIG study were pain, hypertension, an elevation of lipase or gamma-glutamyl transpeptidase levels, and fever. These AEs occurred in 67% of patients [71]. In the phase II study published by Heinrich et al., rash, constipation, fatigue, muscle pain, and headache occurred in at least 40% of patients [70].

5.11. Nilotinib

The safety of nilotinib was evaluated in a phase III study by Reichardt et al. The most common AEs in the nilotinib treatment group were: abdominal pain, nausea, anorexia, fatigue, weakness, and anemia. The most frequent grade 3 and 4 AEs were weakness, increased lipase activity, abdominal pain, vomiting, increased alanine aminotransferase activity, anorexia, anemia, headache, and myalgia. Many adverse reactions of nilotinib, such as skin and subcutaneous tissue disorders, gastrointestinal disorders, musculoskeletal disorders, and general disorders, are mild and can recover without medical intervention. Most of the above-listed grade 3 and 4 adverse events caused by nilotinib should be dealt with by a dose adjustment or treatment interruption [93].

5.12. Crenolanib

The safety of crenolanib was evaluated in a phase II study. The most common grade 3 and 4 AEs included elevated liver function parameters and anemia [74].

6. Fertility and Pregnancy

An additional important topic related to GIST treatment in young people is the impact on fertility and pregnancy. The potential effects of cancer treatments on pregnancy and a patient's future fertility constitute a significant concern. They may affect the quality of life for patients treated due to cancer and cancer survivors.

Table 4 shows the effects of anticancer treatment for GISTs on fertility, pregnancy, and lactation.

Data about the effects of anticancer treatment in GISTs on fertility, pregnancy, and lactation is limited. Wael et al. assessed the effect of imatinib on the placenta and implantation in a mouse model. Significant changes that may determine fetal growth were observed. They found changes in the epigenetic markers of essential genes imprinted in the placenta and a reduction in the labyrinthine zone and blood vessels in the placenta. Moreover, an effect on placental growth was observed in case of treatment discontinuation before pregnancy. This research may indicate that imatinib has a long-term effect on pregnancy and implantation. More extended drug withdrawal before pregnancy or additional monitoring for possible placental failure should be considered [94].

Pye et al. retrospectively analyzed 180 women who became pregnant during treatment with imatinib. They published data from 125 pregnancies: 71% of the women were exposed during the first trimester of pregnancy, and 26% of the women were exposed throughout the pregnancy. Of these 125 pregnancies, 28% resulted in termination of the pregnancy and 15% in spontaneous abortion. In total, 9.6% of the newborns had fetal abnormalities: hydrocephalus, craniosynostosis, hypoplastic lungs, renal agenesis, evisceration, and scoliosis. Some fetal abnormalities were more frequent than expected in the normal population [95]. The possible effects of imatinib at the fetal–maternal and placental interface have not been studied. Epigenetic changes may be hidden after the birth of a child who previously appeared normal, which may have long-term health consequences [95,96].

The use of tyrosine kinase inhibitors (TKIs) during pregnancy is still uncommon.

Imatinib treatment affects sperm survival and activity, as described in a report that included semen samples from 48 men treated with imatinib for CML [97].

Table 4. The effects of anticancer treatment for GISTs on fertility, pregnancy, and lactation.

Drug	Fertility	Pregnancy	Lactation
imatinib [43]	In non-clinical studies, the fertility of male and female rats was not affected, although effects on reproductive parameters were observed. Studies on patients receiving imatinib and its effect on fertility and gametogenesis have not been performed. Patients concerned about their fertility on imatinib treatment should consult with their physician.	There are limited data on the use of imatinib in pregnant women. There have been post-marketing reports of spontaneous abortions and infant congenital anomalies from women who have taken imatinib. However, studies in animals have shown reproductive toxicity, and the potential risk for the fetus is unknown. Imatinib should not be used during pregnancy unless clearly necessary. If used during pregnancy, the patient must be informed of the potential risk to the fetus.	There is limited information on imatinib distribution in human milk. Studies in two breastfeeding women revealed that both imatinib and its active metabolite could be distributed into human milk. The milk plasma ratio studied in a single patient was determined to be 0.5 for imatinib and 0.9 for the metabolite, suggesting the greater distribution of the metabolite into the milk. Considering the combined concentration of imatinib and the metabolite and the maximum daily milk intake by infants, the total exposure would be expected to be low (~10% of a therapeutic dose). However, since the effects of low-dose exposure of the infant to imatinib are unknown, women should not breastfeed during treatment and for at least 15 days after stopping treatment with imatinib.
sunitinib [83]	Based on nonclinical findings, male and female fertility may be compromised by treatment with sunitinib.	There are no studies on pregnant women using sunitinib, and studies in animals have shown reproductive toxicity, including fetal malformations. Sunitinib should not be used during pregnancy or in women not using effective contraception unless the potential benefit justifies the potential risk to the fetus. If sunitinib is used during pregnancy or if the patient becomes pregnant while on treatment with sunitinib, the patient should be apprised of the potential hazard to the fetus.	Sunitinib and/or its metabolites are excreted in rat milk, and it is not known whether sunitinib or its primary active metabolite is excreted in human milk. Because active substances are commonly excreted in human milk and have the potential for severe adverse reactions in breastfeeding infants, women should not breastfeed while taking sunitinib.
regorafenib [51]	There are no data on the effect of regorafenib on human fertility. Results from animal studies indicate that regorafenib can impair male and female fertility.	There are no data on the use of regorafenib in pregnant women. Regorafenib is suspected of causing fetal harm when administered during pregnancy based on its mechanism of action. Regorafenib should not be used during pregnancy unless clearly necessary and after careful consideration of the benefits for the mother and the risk to the fetus.	It is unknown whether regorafenib or its metabolites are excreted in human milk. In rats, regorafenib or its metabolites are excreted in milk. A risk to the breastfed child cannot be excluded. Regorafenib could harm infant growth and development. Breastfeeding must be discontinued during treatment with regorafenib.
ripretinib [98]	There are no data on the effect of ripretinib on human fertility. Based on findings from animal studies, male and female fertility may be compromised by treatment with ripretinib	There are no data on the use of ripretinib in pregnant women. Based on its mechanism of action, ripretinib is suspected of causing fetal harm when administered during pregnancy, and animal studies have shown reproductive toxicity. Ripretinib should not be used during pregnancy unless the woman's clinical condition requires treatment with ripretinib.	It is unknown whether ripretinib/metabolites are excreted in human milk, and a risk to the breastfed child cannot be excluded. Breastfeeding should be discontinued during treatment with ripretinib and for at least one week after the final dose.
avapritinib [99]	There are no data on the effect of avapritinib on human fertility, and no relevant effects on fertility were observed in a rat fertility study.	There are no data on the use of avapritinib in pregnant women, and studies in animals have shown reproductive toxicity. Avapritinib is not recommended during pregnancy and in women of childbearing potential not using contraception. If avapritinib is used during pregnancy or if the patient becomes pregnant while taking avapritinib, the patient should be advised of the potential risk to the fetus.	It is unknown whether avapritinib/metabolites are excreted in human milk, and a risk to newborns/infants cannot be excluded. Breastfeeding should be discontinued during treatment with avapritinib and for two weeks following the final dose.

7. Conclusions

Gastrointestinal stromal tumors are rare in young adults. They may differ from the disease diagnosed in patients above 40 regarding clinical and molecular characteristics. They are more often diagnosed in the emergency setting and are more often wild type, and more frequently harbor mutations other than *KIT/PDGRFA*. The general rules for surgery and systemic therapy are the same as for patients above 40. However, all decisions should include the social and family roles, expected lifetime, quality of life, and plans for having children. Younger patients have fewer comorbidities, fewer contraindications, and surgical and systemic treatment limitations. The treatment strategy should be defined and implemented by the multidisciplinary team in the sites experienced in sarcomas. Molecular testing should be done whenever possible, as the molecular profile may differ from patients above 40 and may influence the choice of systemic therapy. Surgery optimization with the possible use of preoperative treatment to reduce the extension of surgery needs to be considered. Surgery may also be considered for some metastatic GISTs to improve the OS. Participation in clinical trials, especially after the failure of approved systemic therapies, should always be considered.

Author Contributions: M.D.-Ś., A.K., E.B. and P.R. created the idea for the article. M.D.-Ś., A.K. and E.B. conducted the literature search and data analysis and drafted the work. P.R. critically revised and supervised the work. All authors have read and agreed to the published version of the manuscript.

Funding: This work was further supported by Maria Sklodowska-Curie National Research Institute of Oncology subsidy from Ministry of Science and Higher Education.

Conflicts of Interest: Monika Dudzisz-Śledź has received a speaker honorarium from Pierre Fabre, Merck KGaA, Sanofi Aventis, Novartis, and BMS; has received financial support for attending symposia from Novartis; and has consulted for Merck KGaA and Novartis. Piotr Rutkowski has received an educational grant from Pfizer and a speaker honorarium from Novartis, Pierre Fabre, Eli Lilly, Merck, Sanofi, MSD, and BMS, and has served as a member of the advisory board for MSD, Pierre Fabre, Sanofi, Merck, Novartis, and BMS. Anna Klimczak and Elżbieta Bylina have no conflicts of interest that might be relevant to the contents of this manuscript.

References

1. Miettinen, M.; Lasota, J. Gastrointestinal Stromal Tumors. *Gastroenterol. Clin. N. Am.* **2013**, *42*, 399–415. [CrossRef] [PubMed]
2. Janardan, S.K.; Wechsler, D.S. Caught in the In-Between: Challenges in Treating Adolescents and Young Adults With Cancer. *JCO Oncol. Pract.* **2021**, *17*, 299–301. [CrossRef] [PubMed]
3. NCCN Clinical Practice Guidelines in Oncology (NCCN Guidelines) Gastrointestinal Stromal Tumors (GISTs) Version 1. 2022. Available online: https://www.nccn.org/guidelines/guidelines-detail?category=1&id=1507 (accessed on 20 May 2022).
4. Ahmed, M. Recent Advances in the Management of Gastrointestinal Stromal Tumor. *World J. Clin. Cases* **2020**, *8*, 3142–3155. [CrossRef] [PubMed]
5. Rutkowski, P.; Szumera-Ciećkiewicz, A. Gastrointestinal stromal tumours (GIST)—2018. *Oncol. Clin. Pract.* **2018**, *14*, 399–407. [CrossRef]
6. Dudzisz-Śledź, M.; Rutkowski, P. Advances in the Management of Gastrointestinal Stromal Tumors (GISTs). *Nowotw. J. Oncol.* **2020**, *70*, 280–287. [CrossRef]
7. Søreide, K.; Sandvik, O.M.; Søreide, J.A.; Giljaca, V.; Jureckova, A.; Bulusu, V.R. Global Epidemiology of Gastrointestinal Stromal Tumours (GIST): A Systematic Review of Population-Based Cohort Studies. *Cancer Epidemiol.* **2016**, *40*, 39–46. [CrossRef]
8. van Roggen, J.F.G.; van Velthuysen, M.L.F.; Hogendoorn, P. The Histopathological Differential Diagnosis of Gastrointestinal Stromal Tumours. *J. Clin. Pathol.* **2001**, *54*, 96–102. [CrossRef]
9. Abraham, S.C.; Krasinskas, A.M.; Hofstetter, W.L.; Swisher, S.G.; Wu, T.-T. "Seedling" Mesenchymal Tumors (Gastrointestinal Stromal Tumors and Leiomyomas) Are Common Incidental Tumors of the Esophagogastric Junction. *Am. J. Surg. Pathol.* **2007**, *31*, 1629–1635. [CrossRef]
10. Agaimy, A.; Wünsch, P.H.; Hofstaedter, F.; Blaszyk, H.; Rümmele, P.; Gaumann, A.; Dietmaier, W.; Hartmann, A. Minute Gastric Sclerosing Stromal Tumors (GIST Tumorlets) Are Common in Adults and Frequently Show c-KIT Mutations. *Am. J. Surg. Pathol.* **2007**, *31*, 113–120. [CrossRef]
11. Fletcher, J.A.; Rubin, B.P. KIT Mutations in GIST. *Curr. Opin. Genet. Dev.* **2007**, *17*, 3–7. [CrossRef]
12. Lasota, J.; Dansonka-Mieszkowska, A.; Sobin, L.H.; Miettinen, M. A Great Majority of GISTs with PDGFRA Mutations Represent Gastric Tumors of Low or No Malignant Potential. *Lab. Investig.* **2004**, *84*, 874–883. [CrossRef] [PubMed]

13. IJzerman, N.S.; Drabbe, C.; den Hollander, D.; Mohammadi, M.; van Boven, H.; Desar, I.M.E.; Gelderblom, H.; Grünhagen, D.J.; Reyners, A.K.L.; van Noesel, M.M.; et al. Gastrointestinal Stromal Tumours (GIST) in Young Adult (18–40 Years) Patients: A Report from the Dutch GIST Registry. *Cancers* **2020**, *12*, 730. [CrossRef] [PubMed]
14. Kang, G.; Park, Y.S.; Jung, E.-S.; Joo, M.; Kang, M.S.; Ahn, S.; Kang, G.H.; Kim, K.-M. Gastrointestinal Stromal Tumors in Children and Young Adults: A Clinicopathologic and Molecular Genetic Study of 22 Korean Cases. *APMIS* **2013**, *121*, 938–944. [CrossRef] [PubMed]
15. Prakash, S.; Sarran, L.; Socci, N.; DeMatteo, R.P.; Eisenstat, J.; Greco, A.M.; Maki, R.G.; Wexler, L.H.; LaQuaglia, M.P.; Besmer, P.; et al. Gastrointestinal Stromal Tumors in Children and Young Adults: A Clinicopathologic, Molecular, and Genomic Study of 15 Cases and Review of the Literature. *J. Pediatric Hematol. Oncol.* **2005**, *27*, 179–187. [CrossRef]
16. Wada, R.; Arai, H.; Kure, S.; Peng, W.-X.; Naito, Z. "Wild Type" GIST: Clinicopathological Features and Clinical Practice. *Pathol. Int.* **2016**, *66*, 431–437. [CrossRef]
17. Nannini, M.; Urbini, M.; Astolfi, A.; Biasco, G.; Pantaleo, M.A. The Progressive Fragmentation of the KIT/PDGFRA Wild-Type (WT) Gastrointestinal Stromal Tumors (GIST). *J. Transl. Med.* **2017**, *15*, 113. [CrossRef]
18. Pitsava, G.; Settas, N.; Faucz, F.R.; Stratakis, C.A. Carney Triad, Carney-Stratakis Syndrome, 3PAS and Other Tumors Due to SDH Deficiency. *Front. Endocrinol.* **2021**, *12*, 462. [CrossRef]
19. Zhang, L.; Smyrk, T.C.; Young, W.F.; Stratakis, C.A.; Carney, J.A. Gastric Stromal Tumors in Carney Triad Are Different Clinically, Pathologically, and Behaviorally from Sporadic Gastric Gastrointestinal Stromal Tumors: Findings in 104 Cases. *Am. J. Surg. Pathol.* **2010**, *34*, 53–64. [CrossRef]
20. Miettinen, M.; Fetsch, J.F.; Sobin, L.H.; Lasota, J. Gastrointestinal Stromal Tumors in Patients with Neurofibromatosis 1: A Clinicopathologic and Molecular Genetic Study of 45 Cases. *Am. J. Surg. Pathol.* **2006**, *30*, 90–96. [CrossRef]
21. Vincenzi, B.; Nannini, M.; Fumagalli, E.; Bronte, G.; Frezza, A.M.; Lisi, D.D.; Ceruso, M.S.; Santini, D.; Badalamenti, G.; Pantaleo, M.A.; et al. Imatinib Dose Escalation versus Sunitinib as a Second Line Treatment in KIT Exon 11 Mutated GIST: A Retrospective Analysis. *Oncotarget* **2016**, *7*, 69412–69419. [CrossRef]
22. Joensuu, H.; Wardelmann, E.; Sihto, H.; Eriksson, M.; Sundby Hall, K.; Reichardt, A.; Hartmann, J.T.; Pink, D.; Cameron, S.; Hohenberger, P.; et al. Effect of KIT and PDGFRA Mutations on Survival in Patients with Gastrointestinal Stromal Tumors Treated With Adjuvant Imatinib: An Exploratory Analysis of a Randomized Clinical Trial. *JAMA Oncol.* **2017**, *3*, 602–609. [CrossRef] [PubMed]
23. Rutkowski, P.; Ziętek, M.; Cybulska-Stopa, B.; Streb, J.; Głuszek, S.; Jankowski, M.; Łopacka-Szatan, K.; Las-Jankowska, M.; Hudziec, P.; Klimczak, A.; et al. The Analysis of 3-Year Adjuvant Therapy with Imatinib in Patients with High-Risk Molecular Profiled Gastrointestinal Stromal Tumors (GIST) Treated in Routine Practice. *Eur. J. Surg. Oncol.* **2021**, *47*, 1191–1195. [CrossRef] [PubMed]
24. Antonescu, C.R.; Besmer, P.; Guo, T.; Arkun, K.; Hom, G.; Koryotowski, B.; Leversha, M.A.; Jeffrey, P.D.; Desantis, D.; Singer, S.; et al. Acquired Resistance to Imatinib in Gastrointestinal Stromal Tumor Occurs through Secondary Gene Mutation. *Clin. Cancer Res.* **2005**, *11*, 4182–4190. [CrossRef] [PubMed]
25. Astolfi, A.; Pantaleo, M.A.; Indio, V.; Urbini, M.; Nannini, M. The Emerging Role of the FGF/FGFR Pathway in Gastrointestinal Stromal Tumor. *Int. J. Mol. Sci.* **2020**, *21*, 3313. [CrossRef]
26. Hostein, I.; Faur, N.; Primois, C.; Boury, F.; Denard, J.; Emile, J.-F.; Bringuier, P.-P.; Scoazec, J.-Y.; Coindre, J.-M. BRAF Mutation Status in Gastrointestinal Stromal Tumors. *Am. J. Clin. Pathol.* **2010**, *133*, 141–148. [CrossRef]
27. Brenca, M.; Rossi, S.; Polano, M.; Gasparotto, D.; Zanatta, L.; Racanelli, D.; Valori, L.; Lamon, S.; Tos, A.P.D.; Maestro, R. Transcriptome Sequencing Identifies *ETV6-NTRK3* as a Gene Fusion Involved in GIST. *J. Pathol.* **2016**, *238*, 543–549. [CrossRef]
28. Shi, E.; Chmielecki, J.; Tang, C.-M.; Wang, K.; Heinrich, M.C.; Kang, G.; Corless, C.L.; Hong, D.; Fero, K.E.; Murphy, J.D.; et al. FGFR1 and NTRK3 Actionable Alterations in "Wild-Type" Gastrointestinal Stromal Tumors. *J. Transl. Med.* **2016**, *14*, 339. [CrossRef]
29. Fero, K.E.; Coe, T.M.; Fanta, P.T.; Tang, C.-M.; Murphy, J.D.; Sicklick, J.K. Surgical Management of Adolescents and Young Adults with Gastrointestinal Stromal Tumors: A US Population-Based Analysis. *JAMA Surg.* **2017**, *152*, 443–451. [CrossRef]
30. Casali, P.G.; Blay, J.Y.; Abecassis, N.; Bajpai, J.; Bauer, S.; Biagini, R.; Bielack, S.; Bonvalot, S.; Boukovinas, I.; Bovee, J.V.M.G.; et al. Gastrointestinal Stromal Tumours: ESMO–EURACAN–GENTURIS Clinical Practice Guidelines for Diagnosis, Treatment and Follow-Up. *Ann. Oncol.* **2022**, *33*, 20–33. [CrossRef]
31. Chen, K.; Zhou, Y.-C.; Mou, Y.-P.; Xu, X.-W.; Jin, W.-W.; Ajoodhea, H. Systematic Review and Meta-Analysis of Safety and Efficacy of Laparoscopic Resection for Gastrointestinal Stromal Tumors of the Stomach. *Surg. Endosc.* **2015**, *29*, 355–367. [CrossRef]
32. Joensuu, H.; Eriksson, M.; Sundby Hall, K.; Hartmann, J.T.; Pink, D.; Schütte, J.; Ramadori, G.; Hohenberger, P.; Duyster, J.; Al-Batran, S.-E.; et al. One vs Three Years of Adjuvant Imatinib for Operable Gastrointestinal Stromal Tumor: A Randomized Trial. *JAMA* **2012**, *307*, 1265–1272. [CrossRef] [PubMed]
33. Joensuu, H.; Eriksson, M.; Sundby Hall, K.; Reichardt, A.; Hermes, B.; Schuette, J.; Cameron, S.; Hohenberger, P.; Jost, P.; Al-Batran, S.-E.; et al. Three versus One Year of Adjuvant Imatinib for High-Risk Gastrointestinal Stromal Tumor (GIST): Survival Analysis of a Randomized Trial after 10 Years of Follow-Up. *J. Clin. Oncol.* **2020**, *38*, 11503. [CrossRef]
34. Raut, C.P. Adjuvant Therapy in GIST: Yes/No and How Long? In Proceedings of the Society of Surgical Oncology (SSO) 2021 International Conference on Surgical Care, Virtual, 18–19 March 2021.

35. Raut, C.P.; Espat, N.J.; Maki, R.G.; Araujo, D.M.; Trent, J.; Williams, T.F.; Purkayastha, D.D.; DeMatteo, R.P. Efficacy and Tolerability of 5-Year Adjuvant Imatinib Treatment for Patients with Resected Intermediate- or High-Risk Primary Gastrointestinal Stromal Tumor. *JAMA Oncol.* **2018**, *4*, e184060. [CrossRef]
36. Miettinen, M.; Lasota, J. Gastrointestinal Stromal Tumors: Review on Morphology, Molecular Pathology, Prognosis, and Differential Diagnosis. *Arch. Pathol. Lab. Med.* **2006**, *130*, 1466–1478. [CrossRef] [PubMed]
37. Miettinen, M.; Lasota, J. Gastrointestinal Stromal Tumors: Pathology and Prognosis at Different Sites. *Semin. Diagn. Pathol.* **2006**, *23*, 70–83. [CrossRef] [PubMed]
38. NCCN Clinical Practice Guidelines in Oncology (NCCN Guidelines) Adolescent and Young Adult (AYA) Oncology Version 2.2022. Available online: https://www.nccn.org/professionals/physician_gls/pdf/aya.pdf (accessed on 20 May 2022).
39. Rutkowski, P.; Andrzejuk, J.; Bylina, E.; Osuch, C.; Switaj, T.; Jerzak vel Dobosz, A.; Grzesiakowska, U.; Jurkowska, M.; Woźniak, A.; Limon, J.; et al. What Are the Current Outcomes of Advanced Gastrointestinal Stromal Tumors: Who Are the Long-Term Survivors Treated Initially with Imatinib? *Med. Oncol.* **2013**, *30*, 765. [CrossRef]
40. Demetri, G.D.; von Mehren, M.; Blanke, C.D.; Van den Abbeele, A.D.; Eisenberg, B.; Roberts, P.J.; Heinrich, M.C.; Tuveson, D.A.; Singer, S.; Janicek, M.; et al. Efficacy and Safety of Imatinib Mesylate in Advanced Gastrointestinal Stromal Tumors. *N. Engl. J. Med.* **2002**, *347*, 472–480. [CrossRef]
41. Blanke, C.D.; Demetri, G.D.; von Mehren, M.; Heinrich, M.C.; Eisenberg, B.; Fletcher, J.A.; Corless, C.L.; Fletcher, C.D.M.; Roberts, P.J.; Heinz, D.; et al. Long-Term Results from a Randomized Phase II Trial of Standard- versus Higher-Dose Imatinib Mesylate for Patients with Unresectable or Metastatic Gastrointestinal Stromal Tumors Expressing KIT. *J. Clin. Oncol.* **2008**, *26*, 620–625. [CrossRef]
42. Brzozowska, M.; Wierzba, W.; Szafraniec-Buryło, S.; Czech, M.; Połowinczak-Przybyłek, J.; Potemski, P.; Śliwczyński, A. Real-World Evidence of Patient Outcome Following Treatment of Advanced Gastrointestinal Stromal Tumor (GIST) with Imatinib, Sunitinib, and Sorafenib in Publicly Funded Health Care in Poland. *Med. Sci. Monit.* **2019**, *25*, 3846–3853. [CrossRef]
43. EMA Glivec. Available online: https://www.ema.europa.eu/en/medicines/human/EPAR/glivec (accessed on 4 June 2022).
44. Heinrich, M.C.; Jones, R.L.; von Mehren, M.; Schöffski, P.; Serrano, C.; Kang, Y.-K.; Cassier, P.A.; Mir, O.; Eskens, F.; Tap, W.D.; et al. Avapritinib in Advanced PDGFRA D842V-Mutant Gastrointestinal Stromal Tumour (NAVIGATOR): A Multicentre, Open-Label, Phase 1 Trial. *Lancet Oncol.* **2020**, *21*, 935–946. [CrossRef]
45. Jones, R.L.; Serrano, C.; von Mehren, M.; George, S.; Heinrich, M.; Kang, Y.-K.; Schoeffski, P.; Cassier, P.; Mir, O.; Chawla, S.P.; et al. 1621MO Long-Term Efficacy, Tolerability and Overall Survival in Patients (Pts) with Unresectable or Metastatic (U/M) PDGFRA D842V-Mutant Gastrointestinal Stromal Tumour (GIST) Treated with Avapritinib: NAVIGATOR Phase I Trial Update. *Ann. Oncol.* **2020**, *31*, S973. [CrossRef]
46. Bauer, S.; George, S.; Kang, Y.-K.; Tap, W.D.; Zhou, T.; Picazio, N.; Boral, A.L.; Heinrich, M. VOYAGER: An Open-Label, Randomised, Phase III Study of Avapritinib vs Regorafenib in Patients (Pts) with Locally Advanced (Adv) Metastatic or Unresectable Gastrointestinal Stromal Tumour (GIST). *Ann. Oncol.* **2018**, *29*, viii595. [CrossRef]
47. Demetri, G.D.; van Oosterom, A.T.; Garrett, C.R.; Blackstein, M.E.; Shah, M.H.; Verweij, J.; McArthur, G.; Judson, I.R.; Heinrich, M.C.; Morgan, J.A.; et al. Efficacy and Safety of Sunitinib in Patients with Advanced Gastrointestinal Stromal Tumour after Failure of Imatinib: A Randomised Controlled Trial. *Lancet* **2006**, *368*, 1329–1338. [CrossRef]
48. Reichardt, P.; Kang, Y.-K.; Rutkowski, P.; Schuette, J.; Rosen, L.S.; Seddon, B.; Yalcin, S.; Gelderblom, H.; Williams, C.C.; Fumagalli, E.; et al. Clinical Outcomes of Patients with Advanced Gastrointestinal Stromal Tumors: Safety and Efficacy in a Worldwide Treatment-Use Trial of Sunitinib. *Cancer* **2015**, *121*, 1405–1413. [CrossRef]
49. George, S.; Blay, J.Y.; Casali, P.G.; Le Cesne, A.; Stephenson, P.; Deprimo, S.E.; Harmon, C.S.; Law, C.N.J.; Morgan, J.A.; Ray-Coquard, I.; et al. Clinical Evaluation of Continuous Daily Dosing of Sunitinib Malate in Patients with Advanced Gastrointestinal Stromal Tumour after Imatinib Failure. *Eur. J. Cancer* **2009**, *45*, 1959–1968. [CrossRef]
50. Heinrich, M.C.; Maki, R.G.; Corless, C.L.; Antonescu, C.R.; Harlow, A.; Griffith, D.; Town, A.; McKinley, A.; Ou, W.-B.; Fletcher, J.A.; et al. Primary and Secondary Kinase Genotypes Correlate with the Biological and Clinical Activity of Sunitinib in Imatinib-Resistant Gastrointestinal Stromal Tumor. *J. Clin. Oncol.* **2008**, *26*, 5352–5359. [CrossRef]
51. Stivarga. Summary of Product Characteristics. Available online: https://www.ema.europa.eu/en/documents/product-information/stivarga-epar-product-information_en.pdf (accessed on 15 May 2022).
52. George, S.; Wang, Q.; Heinrich, M.C.; Corless, C.L.; Zhu, M.; Butrynski, J.E.; Morgan, J.A.; Wagner, A.J.; Choy, E.; Tap, W.D.; et al. Efficacy and Safety of Regorafenib in Patients With Metastatic and/or Unresectable GI Stromal Tumor After Failure of Imatinib and Sunitinib: A Multicenter Phase II Trial. *J. Clin. Oncol.* **2012**, *30*, 2401–2407. [CrossRef]
53. Demetri, G.D.; Reichardt, P.; Kang, Y.-K.; Blay, J.-Y.; Rutkowski, P.; Gelderblom, H.; Hohenberger, P.; Leahy, M.; von Mehren, M.; Joensuu, H.; et al. Efficacy and Safety of Regorafenib for Advanced Gastrointestinal Stromal Tumours after Failure of Imatinib and Sunitinib (GRID): An International, Multicentre, Randomised, Placebo-Controlled, Phase 3 Trial. *Lancet* **2013**, *381*, 295–302. [CrossRef]
54. Blay, J.-Y.; Serrano, C.; Heinrich, M.C.; Zalcberg, J.; Bauer, S.; Gelderblom, H.; Schöffski, P.; Jones, R.L.; Attia, S.; D'Amato, G.; et al. Ripretinib in Patients with Advanced Gastrointestinal Stromal Tumours (INVICTUS): A Double-Blind, Randomised, Placebo-Controlled, Phase 3 Trial. *Lancet Oncol.* **2020**, *21*, 923–934. [CrossRef]

55. Heinrich, M.C.; Jones, R.L.; Gelderblom, H.; George, S.; Schöffski, P.; von Mehren, M.; Zalcberg, J.R.; Kang, Y.-K.; Abdul Razak, A.R.; Trent, J.C.; et al. INTRIGUE: A Phase III, Randomized, Open-Label Study to Evaluate the Efficacy and Safety of Ripretinib versus Sunitinib in Patients with Advanced Gastrointestinal Stromal Tumor Previously Treated with Imatinib. *J. Clin. Oncol.* **2022**, *40*, 359881. [CrossRef]
56. Blanke, C.D.; Rankin, C.; Demetri, G.D.; Ryan, C.W.; von Mehren, M.; Benjamin, R.S.; Raymond, A.K.; Bramwell, V.H.C.; Baker, L.H.; Maki, R.G.; et al. Phase III Randomized, Intergroup Trial Assessing Imatinib Mesylate at Two Dose Levels in Patients with Unresectable or Metastatic Gastrointestinal Stromal Tumors Expressing the Kit Receptor Tyrosine Kinase: S0033. *J. Clin. Oncol.* **2008**, *26*, 626–632. [CrossRef] [PubMed]
57. Mir, O.; Cropet, C.; Toulmonde, M.; Cesne, A.L.; Molimard, M.; Bompas, E.; Cassier, P.; Ray-Coquard, I.; Rios, M.; Adenis, A.; et al. Pazopanib plus Best Supportive Care versus Best Supportive Care Alone in Advanced Gastrointestinal Stromal Tumours Resistant to Imatinib and Sunitinib (PAZOGIST): A Randomised, Multicentre, Open-Label Phase 2 Trial. *Lancet Oncol.* **2016**, *17*, 632–641. [CrossRef]
58. Rutkowski, P.; Magnan, H.; Chou, A.J.; Benson, C. Treatment of Gastrointestinal Stromal Tumours in Paediatric and Young Adult Patients with Sunitinib: A Multicentre Case Series. *BMC Cancer* **2017**, *17*, 717. [CrossRef]
59. Schlemmer, M.; Heinrich, M.C.; Demetri, G.D. Sunitinib Treatment in Pediatric Patients with Advanced GIST Following Failure of Imatinib. *Pediatric Blood Cancer* **2009**, *52*, 767–771.
60. Park, S.H.; Ryu, M.H.; Ryoo, B.Y.; Im, S.A.; Kwon, H.C.; Lee, S.S.; Park, S.R.; Kang, B.Y.; Kang, Y.K. Sorafenib in Patients with Metastatic Gastrointestinal Stromal Tumors Who Failed Two or More Prior Tyrosine Kinase Inhibitors: A Phase II Study of Korean Gastrointestinal Stromal Tumors Study Group. *Investig. New Drugs* **2012**, *30*, 2377–2383. [CrossRef]
61. Kindler, H.L.; Campbell, N.P.; Wroblewski, K.; Maki, R.G.; D'Adamo, D.R.; Chow, W.A.; Gandara, D.R.; Antonescu, C.; Stadler, W.M.; Vokes, E.E. Sorafenib (SOR) in Patients (Pts) with Imatinib (IM) and Sunitinib (SU)-Resistant (RES) Gastrointestinal Stromal Tumors (GIST): Final Results of a University of Chicago Phase II Consortium Trial. *J. Clin. Oncol.* **2011**, *29*, 10009. [CrossRef]
62. Doebele, R.C.; Drilon, A.; Paz-Ares, L.; Siena, S.; Shaw, A.T.; Farago, A.F.; Blakely, C.M.; Seto, T.; Cho, B.C.; Tosi, D.; et al. Entrectinib in Patients with Advanced or Metastatic NTRK Fusion-Positive Solid Tumours: Integrated Analysis of Three Phase 1–2 Trials. *Lancet Oncol.* **2020**, *21*, 271–282. [CrossRef]
63. Drilon, A.; Laetsch, T.W.; Kummar, S.; DuBois, S.G.; Lassen, U.N.; Demetri, G.D.; Nathenson, M.; Doebele, R.C.; Farago, A.F.; Pappo, A.S.; et al. Efficacy of Larotrectinib in TRK Fusion-Positive Cancers in Adults and Children. *N. Engl. J. Med.* **2018**, *378*, 731–739. [CrossRef]
64. Falchook, G.S.; Long, G.V.; Kurzrock, R.; Kim, K.B.; Arkenau, T.H.; Brown, M.P.; Hamid, O.; Infante, J.R.; Millward, M.; Pavlick, A.C.; et al. Dabrafenib in Patients with Melanoma, Untreated Brain Metastases, and Other Solid Tumours: A Phase 1 Dose-Escalation Trial. *Lancet* **2012**, *379*, 1893–1901. [CrossRef]
65. Falchook, G.S.; Trent, J.C.; Heinrich, M.C.; Beadling, C.; Patterson, J.; Bastida, C.C.; Blackman, S.C.; Kurzrock, R. BRAF Mutant Gastrointestinal Stromal Tumor: First Report of Regression with BRAF Inhibitor Dabrafenib (GSK2118436) and Whole Exomic Sequencing for Analysis of Acquired Resistance. *Oncotarget* **2013**, *4*, 310–315. [CrossRef]
66. Montemurro, M.; Schöffski, P.; Reichardt, P.; Gelderblom, H.; Schütte, J.; Hartmann, J.T.; von Moos, R.; Seddon, B.; Joensuu, H.; Wendtner, C.M.; et al. Nilotinib in the Treatment of Advanced Gastrointestinal Stromal Tumours Resistant to Both Imatinib and Sunitinib. *Eur. J. Cancer* **2009**, *45*, 2293–2297. [CrossRef] [PubMed]
67. Reichardt, P.; Blay, J.-Y.; Gelderblom, H.; Schlemmer, M.; Demetri, G.D.; Bui-Nguyen, B.; McArthur, G.A.; Yazji, S.; Hsu, Y.; Galetic, I.; et al. Phase III Study of Nilotinib versus Best Supportive Care with or without a TKI in Patients with Gastrointestinal Stromal Tumors Resistant to or Intolerant of Imatinib and Sunitinib. *Ann. Oncol.* **2012**, *23*, 1680–1687. [CrossRef] [PubMed]
68. Iclusig. Summary of Product Characteristics. Available online: https://www.ema.europa.eu/en/documents/product-information/iclusig-epar-product-information_en.pdf (accessed on 3 May 2022).
69. Garner, A.P.; Gozgit, J.M.; Anjum, R.; Vodala, S.; Schrock, A.; Zhou, T.; Serrano, C.; Eilers, G.; Zhu, M.; Ketzer, J.; et al. Ponatinib Inhibits Polyclonal Drug-Resistant KIT Oncoproteins and Shows Therapeutic Potential in Heavily Pretreated Gastrointestinal Stromal Tumor (GIST) Patients. *Clin. Cancer Res.* **2014**, *20*, 5745–5755. [CrossRef]
70. Heinrich, M.C.; von Mehren, M.; Demetri, G.D.; Fletcher, J.A.; Sun, J.G.; Kerstein, D.; Conlan, M.G.; George, S. Ponatinib Efficacy and Safety in Patients (Pts) with Advanced Gastrointestinal Stromal Tumors (GIST) after Tyrosine Kinase Inhibitor (TKI) Failure: Results from a Phase 2 Study. *J. Clin. Oncol.* **2015**, *33*, 10535. [CrossRef]
71. Falkenhorst, J.; Hamacher, R.; Reichardt, P.; Ivanyi, P.; Kasper, B.; Hohenberger, P.; Hermes, B.; Kostbade, K.; Pink, D.; Sülberg, H.; et al. Lower-Dosing Ponatinib in Pre-Treated GIST: Results of the POETIG Phase II Trial. *J. Clin. Oncol.* **2020**, *38*, 11536. [CrossRef]
72. Trent, J.C.; Wathen, K.; von Mehren, M.; Samuels, B.L.; Staddon, A.P.; Choy, E.; Butrynski, J.E.; Chugh, R.; Chow, W.A.; Rushing, D.A.; et al. A Phase II Study of Dasatinib for Patients with Imatinib-Resistant Gastrointestinal Stromal Tumor (GIST). *J. Clin. Oncol.* **2011**, *29*, 10006. [CrossRef]
73. Schöffski, P.; Mir, O.; Kasper, B.; Papai, Z.; Blay, J.-Y.; Italiano, A.; Benson, C.; Kopeckova, K.; Ali, N.; Dileo, P.; et al. Activity and Safety of the Multi-Target Tyrosine Kinase Inhibitor Cabozantinib in Patients with Metastatic Gastrointestinal Stromal Tumour after Treatment with Imatinib and Sunitinib: European Organisation for Research and Treatment of Cancer Phase II Trial 1317 "CaboGIST". *Eur. J. Cancer* **2020**, *134*, 62–74. [CrossRef]

74. Heinrich, M.C.; Griffith, D.; McKinley, A.; Patterson, J.; Presnell, A.; Ramachandran, A.; Debiec-Rychter, M. Crenolanib Inhibits the Drug-Resistant PDGFRA D842V Mutation Associated with Imatinib-Resistant Gastrointestinal Stromal Tumors. *Clin. Cancer Res.* **2012**, *18*, 4375–4384. [CrossRef]
75. von Mehren, M.; Tetzlaff, E.D.; Macaraeg, M.; Davis, J.; Agarwal, V.; Ramachandran, A.; Heinrich, M.C. Dose Escalating Study of Crenolanib Besylate in Advanced GIST Patients with PDGFRA D842V Activating Mutations. *J. Clin. Oncol.* **2016**, *34*, 11010. [CrossRef]
76. Blay, J.-Y.; Heinrich, M.C.; Hohenberger, P.; Casali, P.G.; Rutkowski, P.; Serrano-Garcia, C.; Jones, R.L.; Hall, K.S.; Eckardt, J.R.; von Mehren, M. A Randomized, Double-Blind, Placebo-Controlled, Phase III Study of Crenolanib in Advanced or Metastatic GIST Patients Bearing a D842V Mutation in PDGFRA: The CrenoGIST Study. *J. Clin. Oncol.* **2017**, *35*, TPS11080. [CrossRef]
77. Yebra, M. Human Succinate Dehydrogenase-Deficient Gastrointestinal Stromal Tumors Are Seinsitive to Temozolomide via Induction of ER Stress and DNA Damage. In Proceedings of the CTOS, Virtual, 18–21 November 2020.
78. Yebra, M.; Bhargava, S.; Kumar, A.; Burgoyne, A.M.; Tang, C.-M.; Yoon, H.; Banerjee, S.; Aguilera, J.; Cordes, T.; Sheth, V.; et al. Establishment of Patient-Derived Succinate Dehydrogenase-Deficient Gastrointestinal Stromal Tumor Models for Predicting Therapeutic Response. *Clin. Cancer Res.* **2022**, *28*, 187–200. [CrossRef] [PubMed]
79. Kang, Y.-K.; Ryu, M.-H.; Yoo, C.; Ryoo, B.-Y.; Kim, H.J.; Lee, J.J.; Nam, B.-H.; Ramaiya, N.; Jagannathan, J.; Demetri, G.D. Resumption of Imatinib to Control Metastatic or Unresectable Gastrointestinal Stromal Tumours after Failure of Imatinib and Sunitinib (RIGHT): A Randomised, Placebo-Controlled, Phase 3 Trial. *Lancet Oncol.* **2013**, *14*, 1175–1182. [CrossRef]
80. Tonyali, O.; Coskun, U.; Yildiz, R.; Karakan, T.; Demirci, U.; Akyurek, N.; Benekli, M.; Buyukberber, S. Imatinib Mesylate-Induced Acute Liver Failure in a Patient with Gastrointestinal Stromal Tumors. *Med. Oncol.* **2010**, *27*, 768–773. [CrossRef] [PubMed]
81. Teng, J.F.T.; Mabasa, V.H.; Ensom, M.H.H. The Role of Therapeutic Drug Monitoring of Imatinib in Patients with Chronic Myeloid Leukemia and Metastatic or Unresectable Gastrointestinal Stromal Tumors. *Ther. Drug Monit.* **2012**, *34*, 85–97. [CrossRef]
82. Hutson, T.E.; Bukowski, R.M.; Rini, B.I.; Gore, M.E.; Larkin, J.M.G.; Figlin, R.A.; Barrios, C.H.; Escudier, B.; Lin, X.; Fly, K.D.; et al. A Pooled Analysis of the Efficacy and Safety of Sunitinib in Elderly Patients (Pts) with Metastatic Renal Cell Carcinoma (MRCC). *J. Clin. Oncol.* **2011**, *29*, 4604. [CrossRef]
83. Sutent. Summary of Product Characteristics. Available online: https://www.ema.europa.eu/en/documents/product-information/sutent-epar-product-information_en.pdf (accessed on 5 May 2022).
84. Sehdev, S. Sunitinib Toxicity Management—A Practical Approach. *Can. Urol. Assoc. J.* **2016**, *10*, S248–S251. [CrossRef]
85. Valle, J.W.; Faivre, S.; Hubner, R.A.; Grande, E.; Raymond, E. Practical Management of Sunitinib Toxicities in the Treatment of Pancreatic Neuroendocrine Tumors. *Cancer Treat. Rev.* **2014**, *40*, 1230–1238. [CrossRef]
86. Adams, V.R.; Leggas, M. Sunitinib Malate for the Treatment of Metastatic Renal Cell Carcinoma and Gastrointestinal Stromal Tumors. *Clin. Ther.* **2007**, *29*, 1338–1353. [CrossRef]
87. Strumberg, D.; Clark, J.W.; Awada, A.; Moore, M.J.; Richly, H.; Hendlisz, A.; Hirte, H.W.; Eder, J.P.; Lenz, H.-J.; Schwartz, B. Safety, Pharmacokinetics, and Preliminary Antitumor Activity of Sorafenib: A Review of Four Phase I Trials in Patients with Advanced Refractory Solid Tumors. *Oncologist* **2007**, *12*, 426–437. [CrossRef]
88. Chamberlain, F.; Farag, S.; Williams-Sharkey, C.; Collingwood, C.; Chen, L.; Mansukhani, S.; Engelmann, B.; Al-Muderis, O.; Chauhan, D.; Thway, K.; et al. Toxicity Management of Regorafenib in Patients with Gastro-Intestinal Stromal Tumour (GIST) in a Tertiary Cancer Centre. *Clin. Sarcoma Res.* **2020**, *10*, 1. [CrossRef]
89. Krishnamoorthy, S.K.; Relias, V.; Sebastian, S.; Jayaraman, V.; Saif, M.W. Management of Regorafenib-Related Toxicities: A Review. *Ther. Adv. Gastroenterol.* **2015**, *8*, 285–297. [CrossRef] [PubMed]
90. Maggo, S.; Dubey, A.P.; Dhull, P.; Singh, N.K. Avapritinib: Novel Hope for Patients with Metastatic Gist with PDGFRA Exon 18 Mutation. *Int. J. Basic Clin. Pharmacol.* **2020**, *9*, 1175–1179. [CrossRef]
91. Zhou, Y.; Zhang, X.; Wu, X.; Zhou, Y.; Zhang, B.; Liu, X.; Wu, X.; Li, Y.; Shen, L.; Li, J. A Prospective Multicenter Phase II Study on the Efficacy and Safety of Dasatinib in the Treatment of Metastatic Gastrointestinal Stromal Tumors Failed by Imatinib and Sunitinib and Analysis of NGS in Peripheral Blood. *Cancer Med.* **2020**, *9*, 6225–6233. [CrossRef] [PubMed]
92. Cortes, J.; Kim, D.-W.; Raffoux, E.; Martinelli, G.; Ritchie, E.; Roy, L.; Coutre, S.; Corm, S.; Hamerschlak, N.; Tang, J.-L.; et al. Efficacy and Safety of Dasatinib in Imatinib-Resistant or -Intolerant Patients with Chronic Myeloid Leukemia in Blast Phase. *Leukemia* **2008**, *22*, 2176–2183. [CrossRef] [PubMed]
93. Wang, Z.; Jiang, L.; Yan, H.; Xu, Z.; Luo, P. Adverse Events Associated with Nilotinib in Chronic Myeloid Leukemia: Mechanisms and Management Strategies. *Expert Rev. Clin. Pharmacol.* **2021**, *14*, 445–456. [CrossRef]
94. Salem, W.; Li, K.; Krapp, C.; Ingles, S.A.; Bartolomei, M.S.; Chung, K.; Paulson, R.J.; Nowak, R.A.; McGinnis, L.K. Imatinib Treatments Have Long-Term Impact on Placentation and Embryo Survival. *Sci. Rep.* **2019**, *9*, 2535. [CrossRef]
95. Pye, S.M.; Cortes, J.; Ault, P.; Hatfield, A.; Kantarjian, H.; Pilot, R.; Rosti, G.; Apperley, J.F. The Effects of Imatinib on Pregnancy Outcome. *Blood* **2008**, *111*, 5505–5508. [CrossRef]
96. Hensley, M.L.; Ford, J.M. Imatinib Treatment: Specific Issues Related to Safety, Fertility, and Pregnancy. *Semin. Hematol.* **2003**, *40*, 21–25. [CrossRef]
97. Chang, X.; Zhou, L.; Chen, X.; Xu, B.; Cheng, Y.; Sun, S.; Fang, M.; Xiang, Y. Impact of Imatinib on the Fertility of Male Patients with Chronic Myelogenous Leukaemia in the Chronic Phase. *Target. Oncol.* **2017**, *12*, 827–832. [CrossRef]

98. Qinlock. Summary of Product Characteristics. Available online: https://www.ema.europa.eu/en/documents/product-information/qinlock-epar-product-information_en.pdf (accessed on 10 May 2022).
99. Ayvakyt. Summary of Product Characteristics. Available online: https://www.ema.europa.eu/en/documents/product-information/ayvakyt-epar-product-information_en.pdf (accessed on 10 May 2022).

MDPI
St. Alban-Anlage 66
4052 Basel
Switzerland
Tel. +41 61 683 77 34
Fax +41 61 302 89 18
www.mdpi.com

Cancers Editorial Office
E-mail: cancers@mdpi.com
www.mdpi.com/journal/cancers